Andrew Mckie, Barrister at Clerksroom Manchester, is a specialist in claimant and defendant personal injury, travel and housing law practitioner. Andrew is also now co-owner of a Law Firm in Manchester, Barrister-Direct Limited, where he practices in personal injury, travel and housing law including housing disrepair and cavity wall claims.

Ian Skeate, Barrister joined Clerksroom in January 2010 after practising from St Johns Buildings, Manchester since 2003. He has particular expertise in Personal Injury, Professional Discipline and Regulation, Employment and Costs. Ian is based in Manchester but his practice extends throughout the country especially in London.

A Practical Guide to Running Housing Disrepair and Cavity Wall Claims

2nd Edition

A Practical Guide to Running Housing Disrepair and Cavity Wall Claims

2nd Edition

Andrew Mckie,
Barrister at Law, Clerksroom (Manchester),
LLB Hons, PG DIP and
Former Solicitor Advocate,
Personal Injury, Fraud, Credit Hire

Ian Skeate, Barrister at Law, Clerksroom (Manchester),
BA/MA (Cantab), Personal Injury, Costs,
Disciplinary/Regulatory Work

Law Brief Publishing

Published 2018 by Law Brief Publishing, an imprint of Law Brief Publishing Ltd
30 The Parks
Minehead
Somerset
TA24 8BT

www.lawbriefpublishing.com

Paperback: 978-1-911035-70-1

PREFACE

The Small Claims Track change for personal injury cases is likely to come into effect in April or October 2019. This book looks at other areas personal injury practitioners have begun to focus on, including housing disrepair and cavity wall claims, which may not be subject to fixed costs / small claims track costs.

Generally, this book covers issues in relation to housing disrepair claims and cavity wall claims, and, in particular focuses on the day-to-day issues in practice that one encounters with such cases including spotting the 'winners' and 'losers' quickly and efficiently, practical tips for investigation, litigation tactics, as well as a summary of the law, the important cases, and how to run these claims efficiently.

Chapters concerning cavity walls claims cover the nature of the claim and technical background, relevant standards, who to sue, allegations of negligence and breach of duty, disclosure, remedies, expert evidence, causation, quantum and costs. The book also covers running such claims via Damages Based Agreements (DBAs) and looks at the Damages Based Agreement Regulations 2013.

The book has been fully updated from November 2017 to include CIGA claims, problems with experts' reports, personal injury claims, insolvent defendants and insurance issues, as well as common defences and case management issues. The cavity section of the book includes two updated Chapters and three brand new Chapters covering these aspects.

Chapters concerning housing disrepair seek to cover the nature and technical background of these claims, whom to sue, contractual and statutory rights, establishing breach of duty, expert evidence, causation, quantum and costs.

This is a book that any personal injury practitioner, looking to diversify into these areas, should not be without. The purpose of the 2nd edition has been to look cavity wall claims in more detail now the first claims have become litigated and the issues have become clearer on these cases.

Andrew Mckie and Ian Skeate are Barristers at Clerksroom and co-owners of Barrister-Direct Limited – a BSB entity that litigates housing disrepair and cavity wall claims. Credit and thanks is also given to Simon Redfearn for his Technical Chapters from the 1st edition.

Please note this book and notes are not designed to provide legal advice, and you should always consult a qualified Solicitor/Barrister about any specific set of facts.

Andrew Mckie
Ian Skeate
November 2017

CONTENTS

CHAPTER ONE
AN INTRODUCTION TO HOUSING DISREPAIR AND CAVITY WALL CLAIMS

The 23rd February 2017 was a day that will stick in the mind of many Personal Injury Lawyers – the day we found out that the small claims track would increase for RTA claims from 1st October 2018 (now perhaps 2019) to £5,000.00 and that limit would increase to £2,000 for EL/PL claims. There is no doubt that EL/PL Lawyers will survive these changes, as well as PI firms who deal in clinical negligence, dental negligence, disease, travel claims etc. However, for those who work in an RTA firm, these changes are dramatic and will likely mean it is very difficult for RTA Lawyers to run a practice beyond October 2018, without diversifying into new areas of law. One does not pretend that is an easy feat; it requires learning a new area of law, training staff, changing systems, and learning a whole new area of practice – but it can be done.

Housing Disrepair Claims: An Overview

One area that PI Lawyers may look at is housing disrepair claims. Claims for housing disrepair can include compensation for the cost of repairs to the property, inconvenience in respect of the landlord's failure to repair and any disruption caused by the works, as well as any personal injury related conditions such as asthma caused by issues such as damp or mould. The claimant can also claim for a reduction or refund of rent if they haven't been able to use part or all of their home because of the disrepair, as well as a potential claim for damage to belongings or furniture caused, for example, by damp or mould. This is an area that ties in very nicely for PI Lawyers who have experience of claims brought under the Defective Premises Act 1972 as the issues are very similar.

By way of illustration a common disrepair is penetrating damp or rising damp causing condensation and mould growth in a property. This could be caused for example by a leaking roof, gutter or cracked wall, leaking pipes or rotten window frames. Condensation is the most common form of damp in rented properties. Condensation appears when moisture in the air comes into contact with a cold surface, such as a window or a cold wall. Mould can grow on walls and window frames. Severe mould growth can be a health problem for people with asthma or other chest problems.

Housing disrepair claims can be a good source of fee income. If a claim includes an application for an order for repair work it will only be allocated to the small claims track if both: the estimated damages are less than £1,000, and, estimated cost of the repair work is less than £1,000. If either amount is expected to be more than £1,000, the court will normally allocate the case to the fast track. Hence, provided the estimated repairs/damages are more than £1,000, the case will be a fast track case, and will be cost bearing in the normal way.

How complex are they to run?

There is a certain amount of complexity to running a housing disrepair case. The values of the claims may be fairly low but the issues are important to the tenants who, for many years, have often lived with the consequences of poor housing and associated disrepair issues. EL/PL Lawyers will have no difficulty adapting to this work, as it is very much akin to similar issues found in a defective premises claim.

For those looking to diversify and, for those who are new to the area, it is a good idea to obtain a book which deals with the fundamentals and also to go on a full day training course to get to grips with the basics and procedure. The same is dealt with in detail in the author's book and associated training courses throughout 2017 and 2018.

There are some practicalities one needs to deal with however. For example, it is likely that if you are an SRA Regulated firm, under the code of conduct, you will need to ensure that your staff are adequately trained to deal with these cases and they have appropriate expertise. You may also need to tell your professional indemnity insurer if you are going to be undertaking a new area of work. Readers are requested to bear this in mind.

The Pitfalls

Claimant Lawyers running PI cases, will have become accustomed to Qualified One Ways Costs Shifting (QOCS) protection. Housing disrepair claims per se will not be protected by QOCS. QOCS will only likely apply to a claim where there is a personal injury element and will only apply to the Personal Injury element, and not the housing disrepair part of the claim. There is therefore a potential liability for the Defendant's costs. As a result, it is important, that the claimant's lawyer considers LEI/BTE protection in the normal way to fund the case, and if no such policy is available, the claimant may be advised to take a suitable ATE policy to cover the risk. Depending on income and the nature of the case, the claimant may be eligible for help with legal costs under the legal aid scheme. However, it is noted that, from 1 April 2013, legal aid is only available where disrepair is a serious risk to a claimant's or a family member's health and safety, thus, most cases will not benefit from legal aid.

Limitation may also be an issue. In most cases, the claimant will have to take court action within six years. If the court action is based on the landlord not meeting the terms of a tenancy agreement, this is a claim for breach of contract. If the court action is based on negligence or nuisance, the six years starts when the damage to property took place. However, be aware that for personal injury cases, the limit is three years from the date of the injury or date of knowledge and this limit applies to the whole claim. Readers should therefore be very careful, given the varying limitation periods to watch out for.

CFA risk assessments are also key to housing disrepair cases. To justify a case on a CFA, especially if an ATE premium is involved, one needs to make sure that there will be sufficient damages to cover the premium and damages for the client, to justify taking on the case. The court can make the landlord do the repair work by making an order called an order for 'Specific Performance,' or an injunction called a, 'Mandatory injunction.' The court also has the power to make a declaration that a tenant can do the repairs themselves and deduct the cost from future rent. The actual net compensation recovered on a case may therefore only be quite small. If the likely damages recovered for inconvenience are not likely to cover the ATE premium, because the period of disrepair was minor and thus damages for rent reduction, inconvenience will only be a few hundred pounds, it may not be suitable for a CFA. It may be difficult to run a case without ATE due to exposure to defendant's cost and the need to cover the disbursement for the costs of the surveyor which is likely to be between £500-1,000, depending upon the complexity of the case.

Timescales

The pre-action protocol which can be found at:- https://www.justice.gov.uk/courts/procedure-rules/civil/protocol/prot_hou gives the landlord 20 working days to respond to the letter of claim, so the timescales are tight. Anyone who is interested in this area of work should read the protocol, as it sets out in detail the procedure and is a good introduction to this type of work.

Cavity Wall Claims

Another area of practice a number of firms are looking at is cavity wall claims. In November 2016 MP Hywel Williams raised an early day motion that stated:-

> "*That this House notes that cavity wall insulation (CWI) has been installed in millions of homes in successive government-backed schemes, but that in many cases the insulation has failed, acting not as a barrier to keep heat on the inside but as a bridge to allow water penetration from the outside; further notes that the actual number of failures is unknown and that the consequential damp and other related and very serious problems have blighted the lives of householders for years; notes that many victims are older, disabled or vulnerable people; notes that many installation companies have gone out of business and that the industry guarantee scheme has serious*

> *failings, including a defensive and sometimes hostile attitude to claimants and provides a guarantee not compensation; notes that successful extraction of CWI is difficult and sometimes poorly completed, leading to further problems; commends the tireless work of Civalli, the victims support group; and calls on the Government to institute measures to actively identify potential victims, rectify failed installations and provide proper compensation".1*

Practitioners undertaking housing disrepair claims, are now regularly seeing cases of damp and mould caused by failed cavity wall insulation.

<u>What is a cavity wall and how does it fail?</u>

The government website, the energy saving trust, says2:

> *"A cavity wall is made up of two walls with a gap in between, known as the cavity; the outer leaf is usually made of brick, and the inner layer of brick or concrete block. If your house was built after the 1920s, it is likely to have cavity walls. Older houses are more likely to have solid walls.. If you can see the brickwork on the outside of the house, look at the pattern of the bricks. If your home has* cavity walls, *the bricks will usually have a regular pattern.*
>
> *Cavity wall insulation fills gaps often found between external walls in homes built after 1920, keeping the warmth in to save energy. It can also help reduce condensation inside the house if this is a problem stemming from your external walls. It is blown into the cavity from the outside of a house.*
>
> *Every part of the wall must be filled with insulation, so it's important that the installer can reach all your external walls.*
>
> *Your home will usually be suitable for cavity wall insulation if it meets these criteria:*
>
> *Its external walls are unfilled cavity walls.*
> *Your cavity is at least 50mm wide.*

The masonry or brickwork of your property is in good condition.
It is more than 10 years old (most newer houses will have insulation already).
The walls are not exposed to driving rain.

If you have any damp patches on your internal walls then they should not be insulated until the problem is resolved. Speak to a builder who specialises in damp prevention.

If your home's external walls are joined to another house, the installer will need to insert a cavity barrier to contain the insulation so your neighbours aren't affected.

Cavity wall insulation can generally be made from three types of material: mineral wool, polystyrene beads, or foamed insulants.

All three are manufactured according to British standards. Foam insulation systems should be certified by the British Board of Agrément and installed according to strict guidance laid out in the associated certificates.

If your house was built in the last 10 years or so, the walls are probably insulated. To find out whether they are you can do the following:

Ask a registered installer for a boroscope inspection. The installer will drill a small hole in your external wall to see if your walls are hollow or filled. Check with your local authority's building control department.

To insulate your cavity walls, the installer drills small holes around 22mm in size at intervals of around 1m in the outside wall of your home. The installer then blows insulation into the cavity using special equipment. Once all the insulation is in, the installer fills the holes in the brickwork so you'll barely notice them.

If you notice problems with your walls after having cavity wall insu-

lation installed such as damp and mould, you should first call the company who carried out the installation. They should discuss the problem with you and revisit your property to establish whether the original survey or installation work is contributing to the problem. If it is, they should arrange for remedial work to fix the problem or for removal of the insulation if required.

If the installer is not willing to help or no longer exists then contact CIGA *to establish whether you have a 25 year CIGA Guarantee. If you do, they should be able to help resolve the issue under the terms of the guarantee. If you don't have a CIGA guarantee, check your paperwork from time of installation to see if you have an independent insurance backed guarantee which provides a similar level of cover. CIGA also have a contact form to* report bad practice.

The symptoms of damp in a home can be very similar and could be caused by poor or inappropriate insulation, or by other issues such as inadequate home maintenance. Neither your installer nor a guarantee provider will agree to carry out or pay for remedial work if they determine that the problems were not caused by inappropriate insulation, bad workmanship or poor materials. If you believe your installer or guarantee provider is unreasonably refusing to help you, you should follow their complaints procedure.

If you are unable to get help from your installer or a guarantee provider, your final option is to speak to a company that can carry out the necessary work. We recommend that you use a company that is accredited with an appropriate scheme. If you are asking them to remove the insulation, you should use a specialist in cavity wall insulation removal accredited with a cavity wall insulation scheme such as CIGA *or* BBA*".*

What can the client claim for?

A failed cavity wall insulation is where the cavity wall insulation fails and allows water to penetrate between the outer and inner walls, causing issues such as damp and mould in the property. It could be the product was over-packed, under filled, where it has slumped, where areas have been missed, vents have been blocked or obstructions missed.

A claim for a failed cavity wall installation seeks compensation from the installer for negligently installing the cavity wall product in the first place and/or installing the product in an unsuitable property. Which says:- "*Damp could occur in properties as a result of cavity-wall insulation if there is a combination of these factors: Your home is exposed to severe levels of wind-driven rain (zones three or four in our map below) Your home is located in an unsheltered position, e.g. not protected by trees or other buildings The external walls are poorly built or maintained with, for example, cracks in the brickwork or rendering*".*3*

Compensation could include the costs of removal of the product and remedying the costs of decoration / remedial repair work to the property. As damp can lead to mould problems and thus respiratory complaints, there can sometimes be a personal injury claim attached.

<u>Some Pitfalls</u>

Firstly, the claims are costly, and like housing disrepair, if run on a CFA, they will need a suitable ATE product, as the claim will not get QOCS protection. The costs of a surveyor could be £500-1,000. The claimant Solicitor must make sure as above that the cost of removal and repair are over £10,000 to ensure the claim is allocated to the fast track and costs are recoverable. Claims under £10,000 can always be explored on a DBA basis.

The main risk on cavity wall cases is in relation to the installer. It is important to check the company is still trading and/or has insurance before taking the claim on. Even if there is insurance in place it must be ascertained whether that insurance covers remedial works and redecoration or just the removal and reinstallation of the insulation. If the company has gone bust, there is clearly no point in taking the claim on a CFA. There is recourse to CIGA but costs are not recoverable and it is debatable as to how long the scheme will last for.

Conclusions

Housing disrepair and cavity wall claims are an area that can be useful for personal injury Lawyers to explore, and they tie in to personal injury claims where damp and mould is an issue. For those used to dealing with EL/PL cases, this is a straightforward step but RTA Lawyers may need more assistance. There is work out there for lawyers to diversify into post October 2018. Those PI firms that can adapt to these changes will survive and indeed do very well in the new areas.

CHAPTER TWO
HOUSING DISREPAIR: CLAIMS UNDER THE LANDLORD AND TENANT ACT 1985

There are a number of causes of action that arise in relation to housing disrepair claims. One duty and the main one concerned in these claims, arises under section 11 of the Landlord and Tenant Act 1985, which sets out:-

<u>Cause of Action under section 11 of the Landlord and Tenant Act 1985</u>

11 Repairing obligations in short leases.

(1) In a lease to which this section applies (as to which, see sections 13 and 14) there is implied a covenant by the lessor—
(a) to keep in repair the structure and exterior of the dwelling-house (including drains, gutters and external pipes),
(b) to keep in repair and proper working order the installations in the dwelling-house for the supply of water, gas and electricity and for sanitation (including basins, sinks, baths and sanitary conveniences, but not other fixtures, fittings and appliances for making use of the supply of water, gas or electricity), and
(c) to keep in repair and proper working order the installations in the dwelling-house for space heating and heating water.

[F6(1A)
If a lease to which this section applies is a lease of a dwelling-house which forms part only of a building, then, subject to subsection (1B), the covenant implied by subsection (1) shall have effect as if—
(a) the reference in paragraph (a) of that subsection to the dwelling-house included a reference to any part of the building in which the lessor has an estate or interest; and
(b) any reference in paragraphs (b) and (c) of that subsection to an installation in the dwelling-house included a reference to an install-

ation which, directly or indirectly, serves the dwelling-house and which either—
(i) forms part of any part of a building in which the lessor has an estate or interest; or
(ii) is owned by the lessor or under his control.

(1B) Nothing in subsection (1A) shall be construed as requiring the lessor to carry out any works or repairs unless the disrepair (or failure to maintain in working order) is such as to affect the lessee's enjoyment of the dwelling-house or of any common parts, as defined in section 60(1) of the Landlord and Tenant Act M11987, which the lessee, as such, is entitled to use.]

(2) The covenant implied by subsection (1) ("the lessor's repairing covenant") shall not be construed as requiring the lessor—
(a) to carry out works or repairs for which the lessee is liable by virtue of his duty to use the premises in a tenant-like manner, or would be so liable but for an express covenant on his part,
(b) to rebuild or reinstate the premises in the case of destruction or damage by fire, or by tempest, flood or other inevitable accident, or
(c) to keep in repair or maintain anything which the lessee is entitled to remove from the dwelling-house.

(3) In determining the standard of repair required by the lessor's repairing covenant, regard shall be had to the age, character and prospective life of the dwelling-house and the locality in which it is situated.

[F7(3A)
In any case where—
(a) the lessor's repairing covenant has effect as mentioned in subsection (1A), and
(b) in order to comply with the covenant the lessor needs to carry out works or repairs otherwise than in, or to an installation in, the dwelling-house, and
(c) the lessor does not have a sufficient right in the part of the building or the installation concerned to enable him to carry out the

required works or repairs,

then, in any proceedings relating to a failure to comply with the lessor's repairing covenant, so far as it requires the lessor to carry out the works or repairs in question, it shall be a defence for the lessor to prove that he used all reasonable endeavours to obtain, but was unable to obtain, such rights as would be adequate to enable him to carry out the works or repairs.]

(4) A covenant by the lessee for the repair of the premises is of no effect so far as it relates to the matters mentioned in subsection (1) (a) to (c), except so far as it imposes on the lessee any of the requirements mentioned in subsection (2)(a) or (c).

(5) The reference in subsection (4) to a covenant by the lessee for the repair of the premises includes a covenant—
(a) to put in repair or deliver up in repair,
(b) to paint, point or render,
(c) to pay money in lieu of repairs by the lessee, or
(d) to pay money on account of repairs by the lessor.
(6) In a lease in which the lessor's repairing covenant is implied there is also implied a covenant by the lessee that the lessor, or any person authorised by him in writing, may at reasonable times of the day and on giving 24 hours' notice in writing to the occupier, enter the premises comprised in the lease for the purpose of viewing their condition and state of repair"

Some helpful guidance on the operation of section 11 is set out in the Court of Appeal case of **SAMUEL EDWARDS** v **BALADAS KUMARASAMY** [2015] EWCA Civ 20:-

"Thus, if read in the modified way that section 11 (1A) requires, contained in the tenancy provides that there is a covenant by the lessor to keep in repair the structure and exterior of the dwelling-house and the structure and exterior of any part of the building in which the lessor has an estate or interest (including drains, gutters and external pipes). It is important to note that section 11 works by

implying a covenant into a tenancy agreement. ***In other words the tenancy contains a compulsory contractual term".***

I propose first to consider the common law. **The general rule is that a covenant to keep premises in repair obliges the covenantor to keep them in repair at all times, so that there is a breach of the obligation immediately a defect occurs. There is an exception where the obligation is the landlord's and the defect occurs in the demised premises themselves, in which case he is in breach of his obligation only when he has information about the existence of the defect such as would put a reasonable landlord on inquiry as to whether works of repair are needed and he has failed to carry out the necessary works with reasonable expedition thereafter: British Telecommunications plc v Sun Life Assurance Society plc [1996] Ch 69, 78-9.** *Where, therefore, a defect appears in the external cladding of a building that is not demised to the tenant (as in Sun Life) or in a sea wall, possession of which is retained by the landlord (as in Murphy v Hurly [1922] 1 AC 369), the landlord is liable even though he has no notice of the disrepair. The latter case conveniently collects the authorities which provide the original rationale for the rule. The principal reason for it was that the courts took the view that it would be unreasonable to interpret a landlord's covenant as absolute when the landlord "had no means of ascertaining the condition" of what was demised: Makin v Watkinson (1870) LR 6 Ex 25 (Bramwell B); Tredway v Machin (1904) 91 LT 310 (Collins MR). The assumption in the older cases is that the landlord had no right of entry in order to inspect the demised property. This is clear from the judgment of Bramwell B in Makin, and succinctly expressed by Collins MR in Tredway:*

"That rule rests upon the principle that the landlord is not the occupier of the premises, and has no means of knowing what is the condition of the premises unless he is told, because he has no right of access to the demised premises, whereas the occupier has the best means of knowing of any want of repair."

In almost all modern leases the landlord will reserve a right to enter the demised property on notice in order to inspect it; and section 11 (6) of the Landlord and Tenant Act 1985 (like its statutory predecessor) confers a similar right on the landlord. *It has never been suggested, however, that the existence of a right of entry makes any difference to the interpretation of a landlord's repairing covenant. The critical distinction, for the purposes of the common law, is between that which is demised and that which is not.*

There is one further point to make about the common law. Where, as here, there has been an express grant of an easement the grant will carry with it an ancillary right on the part of the dominant owner to carry out repairs on the servient owner's land in order to make the easement effective. Thus in the case of the grant of a right of way the dominant owner is entitled to repair the way: Gale on Easements (19th ed) para 1- 91; Newcomen v Coulson (1877) 5 Ch D 133. The judge took the view that Mr Kumarasamy had no right to "take up or mend the path," but unfortunately no one drew this common law principle to her attention.

Whether the landlord must have notice of the disrepair before his liability arises was the issue in O'Brien v Robinson [1973] AC 912, a case which concerned the statutory predecessor of section 11 of the Landlord and Tenant Act 1985. Mr and Mrs O'Brien were injured when the bedroom ceiling of their flat fell on them, and they brought an action against their landlord for damages for breach of his implied covenant. Lord Diplock gave the leading speech with which Lords Reid, Simon and Cross agreed. Lord Diplock began by looking at the history of covenants which statutes inserted into leases. The first such covenant was imported by an Act of 1909. Lord Diplock said:

"But although created by statute the legal nature of this obligation was contractual. Its characteristics were the same as those of an obligation created by a repairing covenant in a lease. What the statute was providing was that any contract for the letting of premises to

which it applied should be read and given effect to as if it contained an express covenant by the landlord to keep the premises in such a state of repair as would make them reasonably fit for human habitation."

He then referred to subsequent legislation and two cases that had considered it, saying:

"At this stage it is sufficient to say that as I read [the two] cases their ratio decidendi was based upon (a) the contractual nature of the landlord's obligation resulting from the statutory requirement that it should be implied as a term in the contract of letting and (b) the legal characteristics of a repairing covenant by a landlord in a lease or tenancy agreement."

Lastly he came to the statutory predecessor of section 11 of which he said:

"...it has the same essential characteristics: (a) that the landlord's obligation results from a statutory requirement that it should be implied as a term in the contract of letting and (b) that the term to be implied has the legal characteristic of a repairing covenant by a landlord in a lease."

Lord Diplock then turned to consider the common law and concluded that:

"... by 1926 the result of half a century of judicial decision was that it was well established that, at any rate where the state of disrepair was known to the tenant, the landlord's obligation to start carrying out any works of repair did not arise until he had information about the existence of a defect in the premises such as would put a reasonable man upon inquiry as to whether works of repair were needed." (Emphasis added)

He then turned to the question whether this applied to cases where the defect was latent, in the sense that the tenant did not know of it, and held that it did. Two points can be made about Lord Diplock's speech. First, the case before the House was one in which the disrepair was within the demise. Second, Lord Diplock regarded the interpretation of the statutory covenant as being governed by the same considerations as the interpretation of an equivalent covenant at common law. This is how Nourse LJ explained the case in Sun Life, when engaged in the reverse process of applying "the rule in O'Brien v Robinson" to the landlord's repairing covenant in a commercial lease. He also described "the rule in O'Brien v Robinson" as:

"... where a defect occurs in the demised premises themselves, a landlord is in breach of his obligation to keep them in repair only when he has information about the existence of the defect such as would put a reasonable landlord on inquiry as to whether works of repair to it are needed and he has failed to carry out the necessary works with reasonable expedition thereafter" (emphasis added)

This court considered the scope of the rule in the context of the statutory covenant in Passley v Wandsworth LBC (1998) 30 HLR 165. Mr Passley was the tenant of a top floor flat under a secure tenancy granted by Wandsworth. In the course of a cold snap pipework on the roof of the block ruptured; and Mr Passley's flat and some of his possessions were damaged by water.

This court held that because the pipes were outside the demise Wandsworth were liable under the covenant irrespective of notice.

The scope of a landlord's covenant to repair is, at common law, defined by the lease itself. However, the application of the statutory covenant depends on the facts. *Whether something is part of the structure and exterior of a dwelling house does not depend on the extent of the demise, but on the facts. That point was decided by this court in Campden Hill Towers Ltd v Gardner [1977] QB 823. The issue in the case was whether a covenant to pay a service charge was invalidated by the statute; and that in turn involved considering what was the structure and exterior of a flat in a block. The land-*

lords argued that the covenant could not apply to anything that was not demised. Megaw LJ said:

"We do not accept the lessors' contention in so far as it would limit "the structure and exterior of the dwelling house" to that which, in the conveyancing meaning, is included in the particular terms of the demise in the lease. Anything which, in the ordinary use of words, would be regarded as a part of the structure, or of the exterior, of the particular "dwelling house," regarded as a separate part of the building, would be within the scope of paragraph (a). Thus, the exclusion by the words of clause 2 of the underlease of "any part of the outside walls" would not have the effect of taking outside the operation of paragraph (a) that which, in the ordinary use of language, would be regarded as the exterior wall of the flat - an essential integral part of the flat, as a dwelling house; that part of the outside wall of the block of flats which constitutes a wall of the flat. The paragraph applies to the outside wall or walls of the flat; the outside of inner party walls of the flat; the outer sides of horizontal divisions between Flat 20 and flats above and below; the structural framework and beams directly supporting floors, ceilings and walls of the flat."

Key Points:-

a) Section 11 works by implying a covenant into a tenancy agreement. In other words the tenancy contains a compulsory contractual term

b) The general rule is that a covenant to keep premises in repair obliges the covenanter to keep them in repair at all times, so that there is a breach of the obligation immediately a defect occurs.

c) There is an exception where the obligation is the landlord's and the defect occurs in the demised premises themselves, in which case he is in breach of his obligation only when he has information about the existence of the defect such as would put a reasonable landlord on inquiry as to whether works of repair are needed and he has failed to carry out the necessary works with reasonable expediency thereafter: *British Telecommunications plc v Sun Life Assurance Society plc [1996] Ch 69, 78-9.*

d) In almost all modern leases the landlord will reserve a right to enter the demised property on notice in order to inspect it; and section 11 (6) of the Landlord and Tenant Act 1985 (like its statutory predecessor) confers a similar right on the landlord.

e) The scope of a landlord's covenant to repair is, at common law, defined by the lease itself. However, the application of the statutory covenant depends on the facts.

The Meaning of to Keep in Repair

In UDDIN & ANOR v ISLINGTON LONDON BOROUGH COUNCIL (2015), a local authority was liable in damages for breach of its repairing obligations in respect of a flat that had been damaged by rising damp. The court held that, whether or not a damp-proof course had previously been inserted, there was a duty to repair and the existence of damp indicated a failure to repair. The key facts from the Judgment are:-

"In his judgment handed down on 6 May 2014, His Honour Judge Mitchell awarded Mr and Mrs Uddin £14,680 as damages for breach of repairing obligations on the part of their landlord, the London Borough of Islington. With the permission of McFarlane LJ, Islington appeal on limited grounds, permission having been refused in relation to their more comprehensive grounds.

2. It was common ground that Mr and Mrs Uddin were the joint weekly tenants of 116B Englefield Road, London, N1. **Since they were weekly tenants, the tenancy contained the covenants implied by section 11 of the Landlord and Tenant Act 1985. These were, so far as material, covenants by Islington to keep in repair the structure and exterior of the dwelling house**

including drains, gutters and external pipes.

The property compromised in the tenancy was a four bedroomed maisonette in the basement and on the ground floor of a Victorian villa. **The Uddins moved in in October 2004 and started to complain about the damp. The greater part of the damages awarded by the judge related to the damp problems.**

4. Miss Rai on behalf of Islington submits that damp may be one of three types, (1) penetrating damp usually caused by the defective state of some part of the property and thus subject to the repairing obligation, (2) condensation damp generally caused by inadequate insulation, ventilation and/or heating and thus not caused by a defective state of the property which is the subject matter of the repairing obligation, and (3) rising damp which can be caused by an inherent defect in the property, but that remedying the defect would involve the property being wholly different to the property that was demised or caused by a defective damp proof membrane, which is subject to the repairing obligation.

We are concerned with rising damp. I do not, however, consider that Miss Rai's submission about an inherent defect is correct. ***The mere fact that damp is caused by an inherent defect does not of itself absolve the landlord from liability.***

That is demonstrated by the case of Elmcroft Developments Ltd v Tankersley-Sawyer [1986] 1 EGLR 47 referred to in Dowding and Reynolds on Dilapidations at paragraph 12-31. That was a case about a basement flat in a Victorian purpose-built mansion block in London. The flat was let for 16 years less three days on a lease containing a landlord's covenant to maintain and keep the exterior of the building and the roof, the main walls, timbers and drains thereof in good and tangible repair and condition. As constructed, the external and party walls of the flat included what had been intended to be a damp-proof course consisting of slates laid horizontally. However, it was ineffective because it was positioned below ground. The flat suffered from extensive rising damp which affected internal plaster, decorations and wood work.

The question was whether the landlord was liable to install a damp-proof course. Both the County Court and the Court of Appeal said yes. Ackner LJ said:

"I therefore conclude that the learned judge was wholly right in the decision which he made as to the failure by the appellants to comply with the repairing covenant and their obligation in regard to curing the damp by using the only practical method at this price, namely, injecting silicone into the wall. [Counsel] was at one stage prepared to concede that, as the plaster became saturated (which, of course, it was), his clients had the obligation to do the necessary patching -- that is removing -- the perished plaster and renewing it. I am bound to say that concession made the resistance to inserting the damp-proof course a strange one. The damp-proof course, once inserted, would on the expert evidence cure the damp. The patching work would have to go on and on and on, because, as the plaster absorbed (as it would) the rising damp, it would have to be renewed, and the cost to the appellants in constantly being involved with this sort of work, one would have thought, would have outweighed easily the cost in doing the job properly. **I have no hesitation in rejecting the submission that the appellants' obligation was repetitively to carry out futile work instead of doing the job properly once and for all."**

Key Points:-

a) The mere fact that damp is caused by an inherent defect does not of itself absolve the landlord from liability.

b) The landlord must put the premises into repair, even if they were not in good repair at the outset of the tenancy.

Meaning of the Structure and Exterior of the Premises

Under the statute, there has to be some part of the structure/exterior of the premises that are in disrepair. The case of SOUTHWARK LONDON BOROUGH COUNCIL V DENISIA MCINTOSH (2001) illustrates this point. The key facts from the Judgment are:-

> "*The Tenant became the Landlord's secure tenant of a maisonette on two floors at No 8 Cam Court, Gloucester Grove Estate Peckham ("the Property") on the 30th October 1992. The Property comprised four bedrooms, a living room, a kitchen, a bathroom and a downstairs toilet. The Property had been to a degree renovated by the Landlord before the Tenant commenced occupation. But not long thereafter and certainly by the latter part of 1993, the Tenant became aware that the Flat was (in the words of the judge) "afflicted with dampness" which permeated the whole Property.* ***There was a smell and pervading aura of damp and evidence of damp in the form of mould, and this became progressively worse. The damp was such that the Tenant had to move out of the main bedroom and move into a smaller one and for this purpose furnish the smaller one with a new bed. In or about 1994 there was a discreet incident when water began to leak through the kitchen ceiling.***
>
> ***The Tenant repeatedly reported the condition of the Property to the Landlord, who did nothing about it. Notwithstanding (as the judge held) continuing complaints about the damp and the ingress of water, the Landlord did not take the complaint seriously until late in 1998 when steps were set in motion which led to remedying the condition in late 1999.***
>
> ***The Tenant brought this action alleging that the Landlord was liable under the covenant for repair implied in the lease for the adverse consequences of the damp.*** *The Judge awarded by way of general damages in respect of the period of serious discomfort endured by the Tenant the sum of £1,500 for each of the five years between the ends of 1994 and 1999 and by way of special damages*

the sum of £350 for specific damage to goods and the cost of purchasing a new bed occasioned by the damp. On the 9th July 2001 Ferris J. gave permission to appeal.

It is established by the decision of the Court of Appeal in Quick v. Taff Ely BC [1968] QB 809 ("Quick") and succeeding cases that "disrepair" is established for the purposes of section 11 if (and only if) there is established to be physical damage to the structure or exterior of the premises which does not rise by reason of the tenant's default. *A landlord is not liable under the covenant merely because there is the most serious damp in the demised premises. In a case such as the present where the occasion for complaint is damp, the tenant must establish either that the damp arises from a breach of the covenant (i.e. physical damage to the structure or exterior of the premises) or that the damp has itself caused damage to the structure or exterior and that this damage in turn has caused the damp of which complaint is made. The primary issue in this case is whether the Tenant in this case has established either of these propositions."*

Key Points:-

a) Disrepair is established for the purposes of section 11 if (and only if) there is physical damage to the structure or exterior of the premises, which does not rise by reason of the tenant's default.

b) A landlord is not liable under the covenant merely because there is damp in the demised premises. In a case such as the present where the occasion for complaint is damp, the tenant must establish, either that the damp arose from a breach of the covenant (i.e. physical damage to the structure or exterior of the premises) or that the damp itself has caused damage to the structure or exterior and that this damage in turn has caused the damp of which complaint is made.

Meaning of Structure

The issues of what is meant by structure was examined in the case of TANYA GRAND v PARAM GILL (2011). The issues was whether plaster formed part of the structure of the premises The key parts of the Judgment are as follows:-

> ***"In my judgment, assuming that the plasterwork damage identified by Mr Lovatt is correctly characterised as damage to the 'structure' of the flat, it was damage for which Mr Gill was responsible under his repairing obligations under both the tenancy agreement and the provisions in section 11(1) of the Landlord and Tenant Act 1985*** *(see Quick v. Taff Ely Borough Council [1986] QB 809, per Dillon LJ, at 818E). A key question, however, which Mr de Waal very properly raised with us, is whether such plasterwork did form part of the 'structure' of the flat. Only if it did was Mr Gill liable to repair it.*
>
> *The judge may have assumed in paragraph 26 that the plaster formed part of the structure of the flat but I find it difficult to read his last sentence of that paragraph as making a finding to that effect. In* **Quick's case the landlord council conceded for the purposes of the appeal that the plaster in the house was part of its structure (see [1986] QB 809, at 820B and 822G) and so that decision provides no authority on the point.** *In Staves & Staves v. Leeds City Council (1991) 23 HLR 107, another decision of this court, a like concession was also made by the landlord council (see per Ewbank J, at 110, and per Lloyd LJ, at 112). Given the concession, that case also cannot be regarded as authority on the point. In Niazi Services Ltd v. van der Loo [2004] 1 WLR 1254 an issue came before this court as to whether plasterwork forms part of 'the structure' of a dwelling-house within the meaning of section 11 of the 1985 Act. Having recognised it as a difficult question, the court explained in paragraph [7] why it did not propose to answer it.*
>
> *The point was, however, the subject of decision by Mr Recorder Thayne Forbes QC* **in Irvine v. Moran (1992) 24 HLR 1; [1991]**

1 EGLR 261. The lease there in question was one to which section 32 of the Housing Act 1961 applied, but section 11(1) of the 1985 Act is, so far as material, in identical terms.

Before the judge were preliminary issues as to which of several items – including 'internal wall plaster' – were part of 'the structure and exterior of the dwelling house' and so within the landlord's repairing covenants imposed by section 32. Dealing first with the meaning of the word 'structure ... of the dwelling house' in that section, Mr Recorder Thayne Forbes said this: '... I have come to the view that the structure of the dwelling house consists of those elements of the overall dwelling house which give it its essential appearance, stability and shape.

The expression does not extend to the many and various ways in which the dwelling house will be fitted out, equipped, decorated and generally made to be habitable. I am not persuaded by Mr Brock [counsel for *the landlord] that one should limit the expression "the structure of the dwelling house" to those aspects of the dwelling house which are load bearing in the sense that that sort of expression is used by professional consulting engineers and the like; but what I do feel is,* ***as regards the words "structure of the dwelling house", that in order to be part of the structure of the dwelling house a particular element must be a material or significant element in the overall construction.*** *To some extent, in every case there will be a degree of fact to be gone into to decide whether or not something is or is not part of the structure of the dwelling house. It is not easy to think of an overall explanation of the meaning of those words which will be applicable in every case and I deliberately decline to attempt such a definition. I am content for the purposes of this case to say that I accept Mr Brock's submission that "structure of the dwelling house" has a more limited meaning than the overall building itself and that it is addressed to those essential elements of the dwelling house which are material to its overall construction.*

That I think is as far as I am able to go. In coming to that conclu-

sion I have been guided to some extent and have sought to follow the approach suggested by Lord Justice Megaw in Campden Hill Towers Limited v. Gardner [1977] 1 QB 823, in particular the passage that appears at page 834F....'

Having so directed himself as to the 'structure ... of the dwelling house', Mr Recorder Thayne Forbes considered which of the items in issue fell within it. Dealing with 'internal wall plaster', he said this:

*'... As I have said, section 32(1)(a) and the words "**structure of the dwelling house" mean something less than the dwelling house overall and limited to the essential material elements that go to make up the structure of the dwelling house. It seems to me that internal wall plaster is more in the nature of a decorative finish and is not part of the essential material elements which go to make up the structure of the dwelling house**. I therefore hold that internal wall plaster and, for the same reasons, the door furniture do not form part of the structure of the dwelling house, bearing in mind I have held that those words mean something less than the overall construction.'*

In this court's decision in Marlborough Park Services Ltd v. Rowe [2006] EWCA Civ 436; [2006] HLR 30, Neuberger LJ (as he then was), with whose judgment Sedley and Tuckey L.JJ agreed, cited from the first passage that I have quoted from Irvine v. Moran (relating to the phrase 'the structure ... of the dwelling house'), saying of it:

'17. Whilst I accept, as I have emphasised, that words such as "structure" or "main structure" must take their meaning from the particular document, lease or statute in which they are found, and from the surrounding circumstances, and while it can be said that any attempt to define them will, to an extent, raise as many questions as it answers, it seems to me that that is a good working definition to bear in mind, albeit not one to apply slavishly.'

The issue in Marlborough was whether wooden joists in the inter-

mediate floor of a two-story maisonette formed part of 'the main structures of the Property' and the manner of its resolution is of no direct assistance for present purposes.

Mr de Waal, for Ms Grand, submitted that the decision in Irvine v. Moran as to plaster on internal walls not forming part of the structure of a dwellinghouse was in principle wrong. He accepted that, applying Mr Recorder Thayne Forbes's guidance as to the meaning of 'the structure ... of the dwellinghouse', internal plasterwork does not give the house stability. But it does, he said, contribute to its appearance and shape. It is, he said, artificial to regard internal plasterwork, whether on walls or ceilings, as purely decorative. The relevant distinction is, he said, between decoration and fittings on the one hand and everything else making up the dwellinghouse on the other.

I have not found this question a straightforward one, the answering of which was not helped by the fact that there was no expert evidence as to the plasterwork in the flat or plasterwork generally, we have no findings about the matter by the judge and had no argument on it from Mr Gill. Despite these disadvantages, I consider that the court must, however, answer the question.

For myself, whilst I would accept and adopt Mr Recorder Thayne Forbes's observations as to the meaning of 'the structure ... of the dwellinghouse' as providing for present purposes, as Neuberger LJ put it, a good working definition, I am respectfully unconvinced by his holding that the plaster finish to an internal wall or ceiling is to be regarded as in the nature of a decorative finish rather than as forming part of the 'structure'. In the days when lath and plaster ceiling and internal partition walls were more common than now, the plaster was, I should have thought, an essential part of the creation and shaping of the ceiling or partition wall, which serve to give a dwellinghouse its essential appearance and shape. I would also regard plasterwork generally, including that applied to external walls, as being ordinarily in the nature of a smooth constructional finish to walls and ceilings, to which the decoration can then be applied, rather than a decorative finish in itself. I would therefore hold that it is part of the 'structure'. **I would accordingly accept that the wall and ceiling plaster in Ms Grand's flat formed part of the 'structure' of the flat for the repair of which Mr Gill was responsible.**

It follows that I consider that the judge should also have found that the disrepair of the damaged plasterwork meant that Mr Gill was in that respect in breach of his repairing obligations and should have compensated Ms Grand appropriately. I admit to at least some uncertainty as to whether, in arriving at his 'full liability' figure of £6,000 in paragraph 33 the judge was including the plasterwork damage, but I infer that he was; and, consistently with that, Mr de Waal's complaint was that he was wrong to discount the £6,000 figure by as much as 90%.

I propose to proceed on that basis. He should, in my view, therefore have awarded Ms Grand (i) full compensation for the plasterwork damage, and (ii) discounted compensation for the remainder of the damage to which the inadequate heating contributed (there was no challenge to the principle of the latter approach that the judge adopted in paragraph 33)"

Key Points:-

a) Only if the plaster formed part of the structure of the property, would there be liability.

b) The structure of the dwelling house consists of those elements of the overall dwelling house, which give it its essential appearance, stability and shape. The expression does not extend to the many and various ways in which the dwelling house will be fitted out, equipped, decorated and generally made to be habitable.

c) As regards the words, "structure of the dwelling house," in order to be part of the structure of the dwelling house, a particular element must be a 'material or significant element' in the overall construction

d) In *Grand*, the Court said –"*I would accordingly accept that the wall and ceiling plaster in Ms. Grand's flat formed part of the 'structure' of the flat for the repair of which Mr Gill was responsible*".

The Definition of Working Order

In (1) YVONNE O'CONNOR (2) EVON CASSEL (3) JEAN WOODHALL (4) DOREEN WATERS V OLD ETONIAN HOUSING ASSOCIATION LTD (2002), the court offered guidance on the construction of the words "keep in... proper working order" in s.11(1)(b) Landlord and Tenant Act 1985, which were reflected in the wording of a landlord's covenant in relation to water pipes. The Court said:-

> "*This appeal raises a point on the construction of section 11(1)(b) of the Landlord and Tenant Act 1985. It raises a similar point on the construction of clause 2(5) of certain tenancy agreements.* ***The question to which the parties would like an answer is what is involved in the obligation to keep pipes carrying water in "proper working order".*** *The tenancies with which this appeal is concerned are of four flats on the top floor of Eton House in Leigh Road North London. The tenancies commenced on various dates between September 1991 and June 1994. The facts as assumed by the judge are appended hereto. The key points assumed are (i) prior to any of the tenancies commencing in 1986/87 the landlords refurbished Eton House and replaced the water pipework with pipework of a smaller bore; (ii) those smaller pipes carried water successfully to the top floors of the relevant premises for some years; (iii) from the summer of 1992 onwards the smaller pipes did not do so the water pressure having fallen; (iv) the previous pipes would have carried water to the third floor successfully during that period; (v) the landlord could have installed a pump to make the smaller pipes carry the water to the third floor; (vi) from 1998, once the Water Authority*

had constructed a new pumping station, the water was carried successfully to the third floor flats by the smaller pipes.

On 22 August 2000 His Honour Judge Marr-Johnson in the Mayor's and City of London Court held, on the facts as assumed, that there was no breach of the obligation to keep pipes in proper working order. His reasoning in essence was that since the smaller pipes worked perfectly successfully for a period of years, and since nothing had altered except the water pressure, there was no failure to keep in proper working order.

On appeal to Blackburne J however, by a judgment dated 9 February 2001, the judge, again on the assumed facts, allowed the appeal. **He held that the obligation to keep in proper working order was an obligation "to ensure that the pipes were physically or mechanically capable of supplying water to the flats" and he held that on the assumed facts the landlord could have so ensured but did not.**

This is thus a second tier appeal brought with leave of Arden LJ. She gave leave on the grounds that there was a real prospect of success on the issue of statutory construction whether a breach of the obligation imposed by section 11(1)(b) of the Landlord and Tenant Act 1985 can arise where an installation is in mechanical order but fails to carry out its function due to some matter for which responsibility cannot be attributed to the landlord. **She also identified a point of principle – an opportunity for the court to consider its decision in Campden Hill Towers Ltd v Gardner [1977] 1 AER 739 in relation to what was held by the House of Lords in Liverpool County Council v Irwin [1977] AC 239.**

Although there is no doubt that the question of statutory construction and the point of principle do arise in the litigation between the tenants and the landlord, the method chosen for deciding the points has not proved satisfactory for two reasons.

The issue as formulated does not adequately accommodate the pecu-

liar features of a case such as this. It seems to us that three issues can be identified, although these issues overlap to a considerable extent:

i) Does the requirement to keep an installation for the supply of water, gas or electricity 'in proper working order' require the landlord to ensure that the installation is so designed and constructed as to be capable of performing its function at the commencement of the tenancy?

ii) To what extent must the installation be designed and constructed in a manner that will cater for variations in the quantity or character of the water, gas or electricity supplied?

iii) If, because of some change in the quantity or character of the supply, the installation is no longer capable of performing its function, is the landlord obliged to adapt the installation in order to accommodate the change?

The first issue:

11. For the tenants, Mr Staddon argued that the landlord's duty at the commencement of the tenancy was to provide installations for the supply of water gas and electricity that were capable of performing their function of conveying the water gas and electricity received from the undertaker to the appliances which used them in each tenant's flat. In so submitting, he relied principally upon the decision of the House of Lords in Liverpool City Council v Irwin, which dealt with the predecessor to section 11 (section 32 of the 1961 Act). The relevant issue was whether lavatory cisterns, which, because of a design defect were liable to overflow, constituted a breach of the landlord's duty to keep the installations in proper working order. As to this, Lord Edmund-Davies commented at p. 269:

"It is clear that section 32(1)(b) of the Housing Act 1961 imposes an absolute duty upon the landlord "to keep in repair and proper working order the installations in the dwelling house- ..." It could

be said that the opening words ("to keep ...") apparently limit the landlord's obligation to preserving the existing plant in its original state and create no obligation to improve plant which was, by its very design, at all times defective and inefficient. But the phrase has to be read as a whole and, as I think, it presupposes that at the inception of the letting the installation was "in proper working order," and that if its design was such that it did not work "properly" the landlord is in breach.

Bathroom equipment which floods when it ought merely to flush is clearly not in "working order," leave alone "proper" working order (if, indeed, the adjective adds anything). To say that such whimsical behaviour is attributable solely to faulty design is to advance an explanation that affords no excuse for the clear failure "to keep in ... proper working order." Just as badly designed apparatus has been held not of "good construction" (Smith v A Davies & Co (Shopfitters) Ltd (1968) 5 K.I.R. 320, per Cooke J), so in my judgment the landlords here were in breach of section 32(1)(b) by supplying bathroom equipment which, due to bad design, throughout behaved as badly as did the Irwins' cistern. I do not, however, find established any of the other statutory breaches alleged."

Implicit in this passage are two principles which bear on the first issue. **First, something which is not working because it is badly designed is not "in proper working order". Second, if there is an obligation to "keep" in "proper working order" there must by inference be an obligation to supply at the commencement of the tenancy the installation in proper working order.** *It is implicit in the speeches of Lord Cross at p.257G, Lord Salmon at p. 264C and Lord Fraser at p.276C that they agreed with Lord Edmund- Davies.*

Mr Staddon suggested that the approach of Lord Edmund-Davies was supported by a dictum of Lord Esher MR in Proudfoot v Hart (1890) 25 QB 42 at 50 where Lord Esher said:

"***But it has been decided – and, I think, rightly decided – that,***

where the premises are not in repair when the tenant takes them, he must put them into repair in order to discharge his obligation under a contract to keep and deliver them up in repair. If the premises are out of repair at any time during the tenancy the landlord is entitled to say to the tenant, "you have now broken your contract to keep them in repair;"..."

Proudfoot v Hart involved consideration of a tenant's duty to 'keep and deliver up the premises in good tenantable repair'. For the landlord Mr Bhose referred us to a number of authorities which establish that the obligation to keep in repair, whether it falls on the landlord or the tenant, does not of itself require the obligee to remedy defects of design or construction, even if their consequence is that the premises are not reasonably habitable. The duty is to prevent the physical condition of the ill-designed premises from deterioration: Quick v Taff-Ely Borough Council [1986] QB 809; Post Office v Aquarius Properties Ltd [1987] 1 All ER 1055; Southwark London Borough Council v Tanner [2001] 1 AC 1; Lee v Leeds City Council [2002] EWCA Civ 06.

We do not consider that these cases are in point. There is an obvious distinction between the duty to keep 'in repair' and the duty to keep 'in proper working order'. We are concerned with the latter duty. An installation cannot be said to be in proper working order if, by reason of a defect in construction or design, it is incapable of working properly.

Accordingly we have no difficulty in resolving the first issue. **At the commencement of the tenancy the installations for the supply of water, gas and electricity must be so designed and constructed as to be capable of performing their functions.** *The assumed facts of this case demonstrate, however, that the question of whether installations for the supply of water, gas or electricity are capable of performing their function can raise a difficult question as to how this is to be judged. This brings us to the second issue.*

The second issue:

The function of water pipes within a building is to convey within the building the water that is supplied to the building. The assumed facts demonstrate that whether the pipes with which this action is concerned will perform this function to all parts of the building depends upon the pressure under which the statutory undertaker supplies the water to the building. To what extent must the pipes supplied by the landlord be designed and constructed in a manner which will ensure that they perform their function, despite variations in water pressure?

Mr Bhose submitted that if, at the time of installation, the pipes conveyed water to all parts of the building, this demonstrated that they were in proper working order. If at a later date they ceased to do so because of a drop in water pressure, this did not and could not mean that they had ceased to be in proper working order. The fact that they were no longer working properly was attributable, not to any shortcoming in the pipes, but to a shortcoming in the supply, for which the landlord was not responsible.

In support of this submission, Mr Bhose relied upon a passage in the judgment of Megaw LJ in the Campden Hill case, dealing with the proper construction of section 32(1)(b) of the Housing Act 1961. That, as we have said, was the predecessor of section 11 of the Landlord and Tenant Act 1985 with which this case is concerned. But it should be emphasised that the main point in the case related to the physical area in which the installations had to be situated, if they were to be covered by the landlord's obligations. Campden Hill decided that on the proper construction of section 13(1) the installations had to be "in" the dwelling house and it was to reverse that decision that sub-sections 1(A) and 1(B) of section 11 were inserted into section 11 of the 1985 Act by the Housing Act 1988. In that context Megaw LJ by reference to section 32(1) said this at p.745J:

"The lessees contend that, despite the words of the paragraph 'installations in the dwelling-house', [in paragraph (b) of section

32(1)] the paragraph applies to anything outside the flat, the proper functioning of which is required in order to enable the installation within the flat to function as it is intended to do. If it were not for the words 'proper working order' in para (b) it would, we think, be difficult to find any support for the lessees' contention. But the inclusion of those words does provide some possible support. It would, however, produce very odd results if that were so. First, there is nothing in s 32 which requires a lessor to provide such installations. Any such obligation would have to be derived from either non- statutory terms of the lease itself, which would be a matter of contractual negotiation, or from some other statute. Secondly, if 'proper working order' did include, for example, the necessity of a supply of hot water to a radiator, or water to a cistern there would be imposed by statute an absolute obligation, with no qualifications, which in some respects would be quite outside the lessors' control. For example, the central heating boiler in the basement may be operated by gas. The gas supply is cut off by the Gas Board for some reason outside the control of the lessors; or the water supply is cut off, or limited to certain times of the day, by the water authority. The hot water radiator, or the water cistern, while in perfectly good repair and perfectly good 'working order', cannot perform their function. If the covenant has the meaning suggested by the lessees, the lessors are liable for breach of the implied covenant.

In our judgment, the meaning of para (b) is as contended for by the lessors. **The installations in the physical confines of the flat must be kept in repair and capable, so far as their own structural and *mechanical condition is concerned, of working properly. But no more than that. The lessors may be under additional obligations but, if so, they do not arise from this statute."***

Mr Bhose sought, as we understood him, to draw from the above the proposition that whether an installation conveying water, gas or electricity is in good working order falls to be determined without reference to the particular characteristics of the supply, at least if it worked properly having regard to the prevailing characteristics at the time that it was installed.

We do not consider that this proposition can be drawn from Camden Hill. The examples given by Megaw L.J. were of temporary and untoward interruptions to the normal supply. Plainly installations deprived of supply in such circumstances could not be said to be out of proper working order. Those examples afford no assistance in answering the question of the extent to which installations must be able to accommodate alterations in the characteristics of the supply.

Mr Staddon submitted that the duty to keep the pipes 'in proper working order' required the landlord to provide pipes that could perform their function of conveying water to the individual tenants' flats at the pressure at which the undertaker supplied the water from time to time. In support of this submission he referred us to a passage in the judgment of May LJ in Wycombe Health Authority v Barnett (19820 5 HLR 84 at p.91:

"I do not think that a water installation in a cottage ceases to be in proper working order within the statutory provision just because the water in one of the pipes freezes. ... In my opinion the phrase "proper working order" in its context relates to the physical or mechanical condition of the installation as such and involves that it shall be capable of working properly as an installation."

Here again we do not think that the passage affords any assistance in answering the second issue. The requirement that the pipes should be 'capable of working properly as an installation' merely begs the question.

We have concluded that there is no authority which bears on the second issue.

This may explain why the two judges who have so far considered this case came to firm conclusions which were diametrically opposed. In accepting Mr Rundell's submissions (who then appeared for the landlord), HHJ Marr-Johnson held:

"In my judgment, the proper analysis in this case is that the physical condition of the installation was such that it was perfectly capable of working properly as an installation, provided only that water was supplied by the statutory undertaker at a sufficient pressure. I conclude, therefore, that on the agreed facts the claimants are unable to show that these defendants have at any time failed to keep in proper working order the installations for the supply of water at the affected premises, either within 1985, or pursuant to the express terms of the Tenancy Agreement applicable in each case."

In reversing this decision, and preferring the argument of Mr Staddon, Blackburne J. held:

"Putting the matter in my own words, the landlord's obligation under section 11, insofar as it relates to keeping in proper working order the installation for the supply of water, is to ensure that the installations in each flat, and also in the remainder of Eton House so far as they serve directly or indirectly the flats in question, are physically or, if one likes, mechanically capable of supplying water to the flats. That does not merely mean that they should be so capable, given certain minimum water pressures; nor does it mean that the obligation is absolute and unqualified such that the landlord will be in breach even if there is an interruption in water supply for which the landlord is not responsible. It merely means, in my judgment, insofar as there is a supply of water to Eton House, that the installations in the building for the supply of water to each flat must be physically capable of ensuring that a supply of water is maintained."

We have concluded that there is no easy test which answers the second issue. The characteristics of the supply of water, gas and electricity are all capable of varying, whether by accident or design. An obligation which required the landlord to provide installations which would function regardless of the vagaries of supply would be manifestly unreasonable. Equally unreasonable would be an obligation which enabled the landlord to supply installations which would accommodate no changes in the character of supply after the date of installation, even though some variations were reasonably to be anticipated.

Our conclusion is that, **insofar as installations for the supply of water, gas and electricity are concerned an installation will be in proper working order if it is able to function under those conditions of supply that it is reasonable to anticipate will prevail. While the test as formulated may appear to involve uncertainty, we suspect that if the evidence relevant to the test is considered, it may be easier to apply in practice.**

Insofar as installations for the supply of water are concerned, we raised before the hearing the possibility that statutory obligations on water undertakers as to the minimum pressure at which water must be supplied might have some bearing on the issue in this case. There are indeed such obligations, but counsel did not find them easy to assemble or to digest. There are also statutory provisions under which bye-laws can be made imposing obligations on property owners as to the design of installations which are to make use of the water supplied. These also might be of relevance.

Thus it seems to us that the answer to the second issue is likely to depend upon the extent to which supply variations should reasonably be anticipated and provided for. This will depend upon the particular facts. We turn to the third issue.

The third issue:

If, after a tenancy has commenced, a variation occurs to the supply which could not reasonably have been anticipated but which requires some alteration to the installation if the installation is to continue to function properly, is the landlord obliged to make that change or modification? Here again we do not think that it is possible to give a categoric answer. An unanticipated change in the nature of the supply of a utility may occur in a variety of circumstances. **Sometimes the change is imposed deliberately because of some scientific or technical advance – a change in the**

voltage of electrical supply, or the change from coal gas to natural gas are examples. In such circumstances the change is likely to be introduced in a manner and subject to conditions under which it is reasonable to expect customers to modify their installations in order to accommodate the change. In such circumstances, business efficacy would suggest that the landlord's duty to keep installations in proper working order must involve the obligation to make the modifications necessary to enable the installations in question to continue to function.

In other circumstances the change may be forced on the undertaker by some unforeseen event – the collapse of a reservoir or a drought, resulting in a drop in water pressure. If the change is likely to be short–lived, the cost of modifying installations in order to accommodate it may be disproportionate. . Where the changed circumstances are likely to persist for a lengthy period, it may seem wholly unreasonable for a landlord to leave his tenants deprived of satisfactory supply for want of relatively modest expenditure on modifications.

How the test of proper working order is to be applied in such circumstances is a question of great difficulty. We have in this judgment sought to illuminate the issues and to explain why it is that we do not believe that the assumed facts enable them to be answered.

A case such as this cries out for compromise – the more so because the attempt to resolve the issue on the basis of assumed facts has failed. If compromise cannot be reached, then the difficult issues raised must be resolved in the light of findings, or binding agreement, of the material facts,

The appeal must be allowed, because Blackburne J could not have reached the answer that he did on the assumed facts. But the same is true of the decision reached by Judge Marr- Johnson".

Key Points:-

a) At the commencement of the tenancy, the installations for the supply of water, gas and electricity must be so designed and constructed as to be capable of performing their functions.

b) Insofar as installations for the supply of water, gas and electricity are concerned an installation will be in proper working order if it is able to function under those conditions of supply that it is reasonable to anticipate will prevail.

c) Sometimes the change is imposed deliberately because of some scientific or technical advance – a change in the voltage of electrical supply, or the change from coal gas to natural gas are examples. In such circumstances, the change is likely to be introduced in a manner and subject to conditions under which it is reasonable to expect customers to modify their installations in order to accommodate the change. In such circumstances, business efficacy would suggest that the landlord's duty to keep installations in proper working order must involve the obligation to make the modifications necessary to enable the installations in question to continue to function.

What does keep in repair mean?

LILIANA MONTOYA V HACKNEY LONDON BOROUGH COUNCIL (2005), the Court dealt with the standard of repairs required by a landlord. The key passages are:-

"*Section 11 of the Landlord and Tenant Act 1985 provides:*

"*(1) In a lease to which this section applies (as to which, see sections 13 and 14) there is implied a covenant by the lessor-*
(a) to keep in repair the structure and exterior of the dwelling-house (including drains, gutters and external pipes),
(b) to keep in repair and proper working order the installations in the dwelling-house for the supply of water, gas and electricity and

for sanitation (including basins, sinks, baths and sanitary conveniences, but not other fixtures, fittings and appliances for making use of the supply of water, gas or electricity), and
(c) to keep in repair and proper working order the installations in the dwelling-house for space heating and heating water."

4. Section 11(3) states:

"In determining the standard of repair required by the lessor's repairing covenant, regard shall be had to the age, character and prospective life of the dwelling-house and the locality in which it is situated."

"MR JUSTICE ASTILL: The respondent issued a claim against the appellant alleging breach of s.11 of the Landlord and Tenant Act 1985, those breaches claiming disrepair at 44 Petherton House, London N4, where she was a tenant. An expert was engaged by the respondent and another by the appellant. The appellant filed a Defence denying disrepair, and on 22nd November 2002 ***a deputy district judge made an order for a single joint expert to inspect the property and report and a copy of that to be provided to both parties. The single joint expert was to be agreed between the parties.***

As a result of that, Mr Reddin, a chartered surveyor, inspected and reported. He identified a number of defects, but only three now remain in contention, those being the twisting to the door to the hallway, the twisting of the door to the lounge or living room (as it is variously called), and the twisting of metal window frames throughout the flat.

Mr Reddin clearly considered subsection (3), as in a schedule annexed to his report he assessed the twisting of the hall door as being "commensurate with age and type of dwelling, not disrepair"; with reference to the living room door, "commensurate with the age and type of dwelling, not disrepair"; in connection with the casement window to the right of the door, "commensurate with the age and type of dwelling"; and in connection with the left-hand casement and the right-hand window, "commensurate with the age and type of dwelling". **He was therefore expressing his expert opinion that it was the age and type of dwelling that had caused the problems which he found had arisen"**

As I have said and as I wish to emphasise, it cannot be stressed strongly enough that the exercise of a judge's discretion will be interfered with only in exceptional circumstances. **Here the learned judge had before him the evidence of a court-appointed expert whose opinion was clearly expressed in his schedule of repairs. His report with the documents annexed was accepted in evidence and his acceptance of the defects concerning these three items clearly (as the judge by inference accepted) placed them within the exception in s.11(3).** *There was no evidential basis which could justify the learned judge from coming to a contrary conclusion. Accordingly, this appeal must be allowed"*

Key Points:-

a) The standard of repair is to be determined by section 11(3) of the Act.

b) Expert evidence will likely be required to determine the test in Section 11(3)

c) If defects were commensurate with the age and type of property, they would not amount to disrepair.

Knowledge / Notice

A key component of any housing disrepair case is that the landlord must have notice of the defect. Knowledge could be actual knowledge i.e. a report by the tenant, or constructive knowledge, the landlord should have known. Lord Diplock stated on *OBrien v Robinson* that:

"... by 1926 the result of half a century of judicial decision was that

> *it was well established that, at any rate where the state of disrepair was known to the tenant,* ***the landlord's obligation to start carrying out any works of repair did not arise until he had information about the existence of a defect in the premises such as would put a reasonable man upon inquiry as to whether works of repair were needed".***

The following cases are illustrative of these points. In ***DINEFWR BOROUGH COUNCIL V ELIZABETH P A JONES (1987),*** the Court discussed the law surrounding actual knowledge as follows:-

> *"There is no doubt that in the past the defendant had availed herself of that particular route for complaint and had received visits and repairs as a result of the complaints which she had made from time to time.* ***I mention the differences in nomenclature because this case turns upon the question of whether the local authority had notice of the lack of repair to these premises. Reliance' is placed by the defendant upon an inspection of the premises by Mr David Thomas, who was in the Housing and Environmental Health Department and not in the Architectural Services Department, to which it was said the complaints should have been addressed. I deal with that matter later in this judgment***
>
> *The second matter which the Judge found was that Mr David Thomas had inspected the premises on 12th February 1985 in order to ascertain the condition of cleanliness, being the business of the Environmental Health Department, of which, as I have said, Mr Thomas was a member.* **Notwithstanding that he had observed disrepair to the premises , the Judge found that this did not constitute actual knowledge on the part of the respondents, and thus there was no notice for the purposes of Section 11 of the Landlord and Tenant Act 1985.**
>
> ***That notice is required is clear from the decision of the House of Lords in <u>O'Brien & Anor v. Robinson</u> (1973). A.C. 912. At page 921D, Lord Morris of Borth-y-Gest said:***

"On a consideration of the meaning of a lessor's obligation to keep premises in repair there has been scope for much reasonable competitive argument. The various authorities (which I do not propose fully to cite) show that every point of view has been explored. The following are some of the contentions that have been pressed. **On the one hand, it has been said that it would be wholly unreasonable to make a lessor liable for failing to remedy a defect of which he was unaware. So the liability to repair is one that arises only upon notice that there is a need to repair. Where by contract between lessor and lessee there has been a covenant to keep in repair the parties must have intended that the obligation of the lessor would only arise if the lessor had notice of want of repair and a condition or stipulation to that effect should be imported into the contract. The lessee in occupation would be in the best position to know of any state of disrepair.** *On the other hand, it has been said that if a lessor chooses or is required to covenant to keep premises in repair then there is an absolute obligation upon him. Alternatively, even if ordinarily there is no obligation on the part of a lessor until he is told by his lessee of a need for repair, a lessee can only give notice of any condition of which he is aware and accordingly cannot give notice of some unknown or unseen condition or latent defect: if, in these circumstances, the lessee suffers injury by reason of the premises not being in repair liability should rest upon the lessor1'.* **Then, at the conclusion of his speech he said that notice was required. Again, that notice need not necessarily come from the tenant, actual notice will suffice — and I would add to that, from some responsible source".**

Constructive Knowledge

"The appellant says that it is not sufficient to say: well, he was a different department and therefore he is shut off from any responsibility to his employers in the local authority in other departments. He went there specifically to inspect the property, and the purpose of this visit, although it was to assess the cleanliness of the property itself, is immaterial when the fact remains that he was an official of the

Council who had expertise and who was capable of assessing the state of repair. ***If in fact he saw a state of disrepair, that knowledge on his part was to be attributed to the local authority, even though he did not pass it on in the form of an official report.***

Indeed, he visited the property again to see if the state of cleanliness had been remedied on 15th July 1985, on the direct instructions of the Housing Department, and having done that, he submitted a report which included a reference to the lack of repair of the premises. *It was on the basis of that report that the Judge found that the local Authority had-knowledge as of that date as to the lack of repair of the property.*

However, Mr Carrott has sought to go further and say that therefore, in addition, the local authority should be notionally saddled with notice of matters which Mr David Thomas had not himself seen, but which an immediate inspection would have disclosed. I do not accept that submission so far as the local authority are concerned. *Much of the disrepair, which it was suggested Mr David Thomas should have seen, possibly would not have been seen by him because of the dirty state of the premises. I think the local authority in this case are affected only by the actual notice of those natters which Mr David Thomas observed when he went to the premises".*

Approaching this matter from a different angle, if the tenant had written a letter to the Chief Executive and had made complaints about these defects then Mr Cronin, for the local authority, conceded that that would have been notice to the local authority. On the particular facts of this case I fail to see how, when the Chief Executive, in that capacity, receives a report which relates to one of the properties owned by the local authority and for which they have the responsibility, that can be distinguished in terms merely on the basis that the report was for a different purpose. *The report is quite clear: it is specifying defects and indicating that the external doors and a window require immediate attention as do certain ceilings and skirtings. It is not necessary* *for the report to have specified the exact position of the ceiling, skirting, doors or the window. This was sufficient information to put the local authority on enquiry and, for that purpose was, in my view specific notice in accordance with, the requirement of law. 'To that extent I take the view that the learned Assistant Recorder misdirected himself"*

On the question of the significance of the knowledge of disrepair obtained by Mr David Thomas on the occasion of his February visit, I add only a reference to Griffin v. Pillet (1926) 1 K.B. 17, and to the statement of Wright J. at page 23, where he says this:

"The matter might arise if when doing repairs to a house the lessor by his agents, the builders, who are actually in the house, notice some further defect in the structure requiring immediate attention of which the lessee is not aware. In such a case the lessor who, through his builders actually on the premises, is aware of the defect, is not in my opinion entitled to defend himself in an action for breach of covenant for not doing those repairs merely on the ground that he had had no actual notice from the lessee".

Key Points:-

a) The burden of proving notice is upon the tenant.

b) Notice will usually arise from notice of disrepair given by the tenant. It will be useful to obtain disclosure of the defendant's repair and complaints log / copies of any correspondence passing between the claimant and the defendant in order to establish such.

c) Notice must come from a responsible source. As in *Dinefur,* a visit to the property by the defendant's agent/employee would not be enough, the claimant has to show the defect would have been visible and obvious upon that visit.

d) The landlord has only a reasonable time to effect repairs once he has been given notice of the need for them – *CALABAR PROPERTIES LTD V STITCHER*: CA 1983

There is no need for notice of a defect outside the demised premises. In *Edwards v Kumarsamy* (2015) the Court said:-

> *"DDJ Gilman held that the paved area between the front door of the block and the car park was part of the structure or exterior of flat 10; and awarded Mr Edwards £3,750 in damages. On appeal HH Judge May QC reversed his decision on that issue and there is no appeal against that part of her decision. However, a new point was taken before the judge; namely that Mr Kumarasamy was liable under the extended covenant implied into the tenancy by section 11 (1A) of the Landlord and Tenant Act 1985.* ***The judge held that he was not, because it was a precondition to liability that notice of the defect had to be given. For the reasons that follow I respectfully disagree with the judge and would allow the appeal,***
>
> ***The next issue on this appeal is whether, as the judge held, Mr Kumarasamy's liability is conditional on notice having been given or whether, as Mr Edwards contends, liability arises as soon as the disrepair exists.***
>
> ***I propose first to consider the common law. The general rule is that a covenant to keep premises in repair obliges the covenantor to keep them in repair at all times, so that there is a breach of the obligation immediately a defect occurs. There is an exception where the obligation is the landlord's and the defect occurs in the demised premises themselves, in which case he is in breach of his obligation only when he has information about the existence of the defect such as would put a reasonable landlord on inquiry as to whether works of repair are needed and he has failed to carry out the necessary works with reasonable expedition thereafter: British Telecommunications plc v Sun Life Assurance Society plc [1996] Ch 69, 78-9. Where, therefore, a defect appears in the external cladding of a building that is not demised to the tenant***

(as in Sun Life) or in a sea wall, possession of which is retained by the landlord (as in Murphy) was the original rationale for the rule. The principal reason for it was that the courts took the view that it would be unreasonable to interpret a landlord's covenant as absolute when the landlord "had no means of ascertaining the condition" of what was demised: Makin v Watkinson (1870) LR 6 Ex 25 (Bramwell B); Tredway v Machin (1904) 91 LT 310 (Collins MR). The assumption in the older cases is that the landlord had no right of entry in order to inspect the demised property. This is clear from the judgment of Bramwell B in Makin, and succinctly expressed by Collins MR in Tredway:

"That rule rests upon the principle that the landlord is not the occupier of the premises, and has no means of knowing what is the condition of the premises unless he is told, because he has no right of access to the demised premises, whereas the occupier has the best means of knowing of any want of repair."

In almost all modern leases the landlord will reserve a right to enter the demised property on notice in order to inspect it; and section 11 (6) of the Landlord and Tenant Act 1985 (like its statutory predecessor) confers a similar right on the landlord. It has never been suggested, however, that the existence of a right of entry makes any difference to the interpretation of a landlord's repairing covenant. **The critical distinction, for the purposes of the common law, is between that which is demised and that which is not"**

*"This court considered the scope of the rule in the context of the statutory covenant in Passley v Wandsworth LBC (1998) 30 HLR 165. Mr Passley was the tenant of a top floor flat under a secure tenancy granted by Wandsworth. In the course of a cold snap pipework on the roof of the block ruptured; and Mr Passley's flat and some of his possessions were damaged by water. **This court held that because the pipes were outside the demise Wandsworth were liable under the covenant irrespective of notice.***

The scope of a landlord's covenant to repair is, at common law, defined by the lease itself. However, the application of the statutory covenant depends on the facts. Whether something is part of the structure and exterior of a dwelling house does not depend on the extent of the demise, but on the facts. That point was decided by this court in Campden Hill Towers Ltd v Gardner [1977] QB 823. The issue in the case was whether a covenant to pay a service charge was invalidated by the statute; and that in turn involved considering what was the structure and exterior of a flat in a block. The landlords argued that the covenant could not apply to anything that was not demised. Megaw LJ said:

"We do not accept the lessors' contention in so far as it would limit "the structure and exterior of the dwelling house" to that which, in the conveyancing meaning, is included in the particular terms of the demise in the lease. Anything which, in the ordinary use of words, would be regarded as a part of the structure, or of the exterior, of the particular "dwelling house," regarded as a separate part of the building, would be within the scope of paragraph (a). Thus, the exclusion by the words of clause 2 of the underlease of "any part of the outside walls" would not have the effect of taking outside the operation of paragraph (a) that which, in the ordinary use of language, would be regarded as the exterior wall of the flat - an essential integral part of the flat, as a dwelling house; that part of the outside wall of the block of flats which constitutes a wall of the flat. The paragraph applies to the outside wall or walls of the flat; the outside of inner party walls of the flat; the outer sides of horizontal divisions between Flat 20 and flats above and below; the structural framework and beams directly supporting floors, ceilings and walls of the flat."

Repairs

Once the notice is given, the landlord has a statutory right to enter, at a reasonable time of day, to view the condition and state of repair giving the tenant 24 hours notice in writing. The repairs must then be done within a reasonable time. "Reasonable time" will be judged by the type and extent of the disrepair. The bigger and more expensive the repair necessary, the longer the time period would probably be. Note however that where the repair necessarily is in the nature of an emergency "reasonable time" may be very short indeed regardless of any other considerations.

Personal Injury Claims

Section 11(1) of the Landlord & Tenant Act 1985 sets out as follows:-

> *"In a Lease to which this section applies, as to which see sections 13 and 14, there is an implied covenant by the Lessor:–*
>
> *(a) To keep in repair the structure and exterior of the dwelling house (including drains, gutters and external pipers.*
>
> *(b) To keep in repair and proper order the installations in the dwelling house for the supply of water, gas and electricity and for sanitation (including basins, sinks, baths and sanitary conveniences, but not other fixtures, fittings and appliances for making use the supply of water, gas and electricity).*
>
> *(c) To keep in repair and proper working order the installations in the dwelling house for space, heating and heating water".*

Morgan –v- Liverpool Corporation 1927 2KB 131 held that for a section 11 implied covenant to ground an action against the landlord, the landlord **must** have notice of the defect to be remedied.

It is clear that only a tenant could bring an action in damages for injury under the legislation, since other visitors to the property, would not have privity of contract to sue under the Lease. An action under the Landlord and Tenant Act 1985, thus presents the claimant with another potential avenue for damages in the event of the tenant being injured in the property and where there are difficulties with any claim under the Defective Premises Act 1972.

CHAPTER THREE
HOUSING DISREPAIR: PERSONAL INJURY CLAIMS AND DISREPAIR UNDER THE DEFECTIVE PREMISES ACT 1972

Section 4 – Personal Injury Claims

The Defective Premises Act 1972 represents an important piece of legislation when dealing with slipping and tripping claims that may occur on a landlord's premises.

Claimant lawyers will find that the Act is particularly useful in dealing with cases where it is likely to be difficult establishing liability under the Occupiers' Liability Act 1957, because the landlord is not the occupier of the property. It is therefore an important cause of action because any claimant who is injured in the landlord's property can bring an action under the Defective Premises Act 1972, and it is not just limited to tenants of the property.

The Duty of Care Under the Act

The duty of care is set out in Section 4 of the Act which says as follows:-

> "*Where premises are let under a tenancy which puts on the landlord an obligation to the tenant for the maintenance or repair of the premises, the landlord owes a duty to all persons who might be reasonably expected to be affected by the defect and the state of the premises a duty to take care as is reasonable in all the circumstances to see that they are reasonably safe from personal injury or from damage to their property caused by the relevant defect.*
>
> *The said duty is owed if the landlord knows (whether as a result of being notified by the tenant or otherwise) or if he ought to have*

> *known in all the circumstances of the relevant defect.*
>
> *In this section relevant defect means a defect in the state of the premises existing at or after the material time or arising from, or continuing because of, an act or omission by the landlord which constitutes or would if he had been on notice of the defect, constituted a failure by him to carry out his obligation to the tenant for the maintenance or repair of the premises and for the purpose of the foregoing provision at the material times means:-*
>
> (a) *When the tenancy commenced before the act, the commencement of the act; and*
>
> (b) *In all other cases, the earliest of the following times that is to say –*
>
> i) *The time that the tenancy commences.*
>
> ii) *The time when the tenancy agreement is entered into.*
>
> iii) *The time when possession is taken of the premises in contemplation of the letting"*

It goes on to say at subsection 4:-

> "*Where premises are let under a tenancy which expressly or impliedly gives the landlord the right to enter into the premises to carry out any description of maintenance or repair of the premises, then, as from the time he first is, or by notice or otherwise can put himself, in a position to exercise the right and so long as he does or can put himself in that position, he shall be treated for the purpose of subsection 1 – 3 but for no other purpose, as if he were under an obligation to the tenant for the description or maintenance of the repair of the premises, but the landlord shall not owe the tenant any duty by virtue of the subsection in respect of any defect in the state of the premises arising from, or continuing because of, a failure to carry out an annex obligation expressly imposed by the tenancy".*

The Relevant Defect

Within the meaning of the Act, the relevant defect is likely to mean a defect that arises from or continues because of an act or omission by the landlord. In the context of slipping and tripping claims this is likely to mean defects, such as leaking water pipes, or, for example, cases involving broken bannisters on stairways which have caused falls down the stairway or carpet or flooring which has not been attached to the floor properly which has caused tripping incidents.

In relation to any case brought under the Act, it will be necessary firstly for the claimant solicitor to obtain a copy of the Lease, to see if the landlord has a repairing obligation under the tenancy. Under normal circumstances the landlord will have the repair obligation, but it will be necessary to check in the Lease. It is noteworthy, that the landlord is not under any duty under the Act in respect of any defect in the premises which arises from or continues because of a failure to carry out an obligation which the tenancy expressly places on the tenant.

This therefore means, that if part of the repair obligation in the property is the tenant's obligation and the tenant has failed to undertake his obligations under the tenancy the landlord would not be liable under the Act.

Therefore it is important for the claimant lawyer to have disclosure of the tenancy agreement at the earliest possible opportunity so the claimant's solicitor can establish exactly what the landlord's repair obligation was under the tenancy.

Is A Landlord An Occupier?

Landlords are not usually occupiers. A landlord parts control of the property by granting a Lease over it. However, a landlord will usually be an occupier for any areas not demised by the Lease i.e. common stairways and halls etc.

Therefore, if the claimant's solicitor is looking at an accident which has occurred in a common part which more often than not will be common stairways or halls, for example in a block of flats or a shared stairway to a property, it is important to check the Lease to consider which areas are demised by the Lease and which are not.

If a landlord still has control of an area not demised by the Lease such as a common stairway or hall, then the landlord would still be the occupier for the purpose of the Occupiers' Liability Act 1957 and a claim could still be brought by the tenant/claimant within the meaning of the Act.

Tenants As Occupiers

If a claimant has slipped or tripped and is not the tenant within the property, then it will be important for the claimant's solicitor to consider not only a potential claim under the Defective Premises Act 1972 against the landlord, but also a claim against the tenant under the Occupiers' Liability Act 1957.

If a landlord has granted a tenancy, then it is likely that the tenant will be the occupier for the purpose of the 1957 Act and, providing that the claimant is a visitor within the meaning of the 1957 Act, in any area demised by the lease, then a claim may be appropriately brought against the tenant under the 1957 Act. It will of course be important in such circumstances to check that the tenant is insured for such a risk or that the tenant has the money to fund any Judgment before such a claim is to be pursued.

Can the Landlord Delegate Liability Under The Defect Premises Act 1972?

Under the Defective Premises Act 1972, the Landlord cannot delegate responsibility under the Act to a letting agent or the like, to carry out the landlord's obligations under the Act.

Whereas the Occupiers Liability Act 1957 has specific provisions in relation to independent contractors and there is a defence under the

Occupiers' Liability Act 1957 where the occupier has employed an independent contractor, no such provisions are drawn into the 1972 Act. It is argued that there is good reason for this, since the Defective Premises Act is designed to protect tenants and other visitors to the premises against unscrupulous landlords who do not carry out their repair obligations properly within the meaning of the Act.

To allow a landlord to delegate such obligations to a letting agent or the like, would frustrate the intention of the Act. It is therefore argued that if a landlord has delegated his responsibilities under the 1972 Act to a letting agent or the like and the letting agent has failed in their obligations under the Act, then the landlord's remedy is to sue the letting agent for breach of contract. The Claimant clearly would have no such cause of action, since they would not be privy to the contract between the letting agent and the landlord. In such circumstances, and one often sees these types of defences in practice given that many landlords now use letting agents to carry out their obligations to tenants, the defendant should be advised that their correct course of action if the matter is to be pursued, is to bring in the letting agent as a part 20 defendant or for indemnity or contribution. It is noted that despite this, the landlord will still be primarily liable under the 1972 Act, providing that the claimant can satisfy the conditions of section 4 of the Act.

<u>Notice of The Relevant Defect</u>

A useful case for any practitioner dealing with a Defective Premises Act case is *Patrick Joseph Hannon –v- Hillingdon Homes Limited 2012*. The claim arose out of an injury to Mr Hannon who suffered when he fell whilst carrying out maintenance work to the central heating and hot water boiler at a house owned by the defendant on 18 February 2008.

The claimant brought his claim in negligence and under the Defective Premises Act 1972. The defendant's main defence to the claim was that it was not negligent and it had a defence to the claim under the Defective Premises Act 1972. The defendant was therefore required to look at the following issues when considering whether there was a claim under the Act:-

a) Were the bannisters part of the furniture?

b) Was their failure to replace the bannisters a relevant defect?

c) Was the defendant liable to the claimant given that the tenant removed the bannisters?

d) Did the defendant have notice of the defect?

The court found in summary that there was a relevant defect in the house that caused the injury to the claimant, it made no difference to the existence of liability that the tenant caused the relevant state of disrepair in the premises. The important issue decided in this case was whether the defendant had notice of the defect.

The defendant in this case argues as follows:-

> "*It was a precondition of Hillingdon's liability under the Defective Premises Act that it had been given notice by the tenant by a notice complied with under the terms of a Lease namely a report which had been given immediately to us as our agent, since it was a defect in the property which was likely to affect public health or stability of the structure. Alternatively it was within the requirement of other faults or disrepair that it had to be reported as soon as possible. He also contended that Hillingdon had never been given notice at best the employees of independent contractors had become aware of the defect when making any of the maintenance visits to the property. That however did not constitute knowledge acquired by Hillingdon*".

The claimant argued against this as follows:-

> "*(a) The notice envisaged by clause 5.10 of the Lease including a report to our agents. Any employers including a self-employed employee, of a contractor visiting the property for maintenance and repair purposes that were being undertaken by the landlords clearly*

such an agent for the purpose of receiving a report of the state of repair of the structure. There have been innumerable such visits over the years, any one of which gave rise to a report since the absence of the bannister were obvious to all who visited the property.

(b) The notice did not have to be in writing. The tenant by giving access to the house to the maintenance engineer or similar invite he was giving him notice of absence of the bannisters. It followed that the tenant or on innumerable occasions giving Hillingdon notice of the defect.

(c) In any event, it was clear that a direct employee for Hillingdon had on occasions, visited the property the most recent of each visit having occurred a few days before Mr Hannon had himself visited. Each visit clearly involved notice being given to Hillingdon.

Hillingdon must have undertaken a number of inspections of the interior of the property as their Landlord in order to ensure that there were no relevant defects that required it to act in order to comply with the remedial and maintenance obligations under the Lease. There was no such evidence that such visits had occurred and if that was so, that absence of inspection visits were itself a breach by Hillingdon as Hillingdon's repair obligations gave rise to an obligation to repair such disrepairs as would have been observed had an inspection visit occurred.

The Court found:-

"Finally, in any event section 4(4) of the DPA gave rise to a liability whereas in this case the Landlord had an express right to enter the property to carry out repairs and a further express right to visit and inspect the property for the purpose of seeing whether it could exercise that right. That right arose even if Hillingdon had not in fact exercised its right to repair and inspect since the DPA liability arose "from the time when the notice or otherwise it could have put itself in a position to exercise" the right to inspect and carry out the repairs. Since Hillingdon could have inspected and repaired the

> *bannister and could have done so at any time over the period of the tenancy, it was liable for not carrying out those repairs even if it did not have notice of the defect. For all of those reasons Hillingdon have been given sufficient and appropriate notice even if, which was not the case, it was necessary for that notice to have been given in order to have found liability under the DPA".*

It can therefore be seen from the case of *Hannon –v- Hillingdon*, that this is an important case in relation to whether actual notice of the defect is required under the Act for the landlord to be liable. To succeed under the Act, the defendant must either therefore have actual or constructive knowledge of the defect under section 4 (2), or, imputed knowledge under section 4 (4) of the Defective Premises Act 1972.

Thus, under section 4 (4) of the Act, the defendant will be under a duty if he had the right or obligation to enter that part of the premises to repair the defect in question. If the defendant did have the right he is under a duty in respect of that defect notwithstanding that he either did not know about it nor could reasonably have known. He is treated for the purpose of the Act as if he were under a duty imposed in respect of the defect he knew or should have known about. Knowledge of the defect is thus in effect implied against the landlord.

Practitioners will note that the majority of tenancy agreements contain the provision for the landlord to enter the premises to carry out the repairs, although this should be checked and a copy of the Lease should be requested.

This also sits squarely with the case of *Sykes –v- Harry 2001 EWCA Civ 167* where the Court said "*it was not necessary to prove that the landlord had notice, actual or constructive, of the defect which gave rise to the risk of injury*".

The court went on to say that this was, "*because the relevant defect is also defined in terms of the landlord's failure to carry out such obligations, and thus the argument might otherwise arise as to whether or not he had notice of the relevant defect, the relevant defect is defined in terms which precludes such arguments".*

In order to establish breach of duty against the landlord, the claimant will still have to show that there was a relevant defect in the premises, and that there was a foreseeable risk of injury in order for the claim to succeed.

Therefore, the claimant will also have to show on balance, that the landlord had the repairing obligation for the property and the right to enter the property for maintenance and repairs under the Lease.

Defective premises claims can be incredibly technical and counsel's advice should always be sought where there is doubt over any provisions of the Act.

<u>Claims for Disrepair under Section 4</u>

A landlord's duty under the Defective Premises Act 1972 s.4 did not extend to putting the demised premises into a safe condition. The issue was explored in *ALKER v COLLINGWOOD HOUSING ASSOCIATION (2007)* and sets out in the Judgement:-

> *"It can be seen that the duty under section 4(1) arises if and only if the following conditions are fulfilled:*
>
> ***1. The landlord owes an obligation to the tenant under the tenancy for the maintenance or repair of the premises – section 4(1).***
>
> ***2. The landlord knew or ought to have known of whatever is the "relevant defect" – section 4(2).***
>
> ***However those requirements are qualified by section 4(4): the landlord is treated as under a section 4(1) duty if he can exercise a right enjoyed by him to enter the premises in order to carry out works of maintenance or repair. The duty itself, however, is only to take reasonable care to protect potentially affected persons***

from injury or damage caused by a "relevant defect". That is defined by section 4(3). I repeat the definition for convenience:

"... a defect in the state of the premises ... arising from, or continuing because of, an act or omission by the landlord which constitutes or would if he had had notice of the defect, have constituted a failure by him - that is a failure by him to carry out his obligation for maintenance or repair".

Here it is common ground that the conditions are met. The appellant owed an obligation for maintenance or repair (clause 3.1 of the Tenancy Agreement). Section 4(2) (the second condition) was not fulfilled because the appellant had no notice of the putative defect, but that omission is repaired by the application of section 4(4) which, as I have said, qualifies the conditions. The appellant had a right of entry for the purpose of repair or maintenance by force of clause 3.12. Accordingly, the only question in the appeal is whether the state of the glass panel constituted a "relevant defect". The appellant, by Mr Underwood QC, said that it did not because the glass panel was in no need of repair or maintenance. The respondent, by Mr Hartley, says that it did because it was dangerous. That antithesis articulates in essence the short and important point on which the appeal turns"

The Recorder essentially adopted the approach urged by the respondent. He said this at paragraph 74:

"The question of fact is whether the replacement of the door or glass panel was something which ought to have been done 'in the interests of all persons who might be expected to be affected by the defect' or 'to remedy any defects which might expose visitors to the premises (or the tenant herself) to the risk of injury.' In ***other words, did it pose a reasonably foreseeable risk of injury?"***

In arriving at this formulation the Recorder purported to draw on the reasoning of this court in McAuley v Bristol City Council [1992] QB 134. That case concerned a garden step which was in disrepair and was unstable. The plaintiff tenant fell on it and broke her ankle. In her subsequent claim she relied on section 4 of the Act.

7. The court's lengthy reasoning was much taken up with the question whether the landlord had impliedly reserved a right to carry out repairs to remove a risk of injury. In the circumstances the court held that such a right was reserved.

Ralph Gibson LJ said this (McAuley v Bristol City Council [1991], [1992] QB 134 para 47):

"I think that, to give business efficacy to the agreement, as Somervell LJ put it in Mint v Good, a right should be implied in the council to carry out repairs for the removal of that risk of injury. A reasonable tenant could not sensibly object to such a right. If the council became aware of a dangerous defect in the steps of a steep garden, as in this case, and asked the tenant for access to repair it, in the interest of all persons who might be expected to be affected by the defect, the court could, in my judgment, properly require the tenant to allow such access on the basis of an implied right in the council to do the work. So limited, I would hold that the implied right to enter to do the necessary repair was proved and the appeal should be dismissed."

Neill LJ said in WLR 984H:

"In my judgment, in order to give business efficacy to this tenancy agreement, the landlord had an implied right to enter the premises to carry out repairs to remedy any defects which might expose visitors to the premises (or the tenant himself) to the risk of injury."

9. Before considering the effect of McAuley and the Recorder's reliance on it I should refer to some other authority. The Recorder referred more than once to Quick v Taff Ely Borough Council [1986] 1 B 809. There this court was concerned with the extent of liability under a repairing covenant implied by section 32(1) of the Housing Act 1961, predecessor to the

Landlord and Tenant Act 1985, section 11. Mr Underwood submits that the court's approach casts light on the sense to be attributed to "relevant defect" in section 4 of the Act. Dillon LJ said at page 818D:

"In my judgment, the key factor in the present case is that disrepair is related to the physical ***condition of whatever has to be repaired, and not to questions of lack of amenity or inefficiency****. I find helpful the observation of Atkin LJ in Anstruther-Gough-Calthorpe v McOscar [1924] 1 KB 716, 734 that repair 'connotes the idea of making good damage so as to leave the subject so far as possible as though it had not been damaged.'* ***Where decorative repair is in question one must look for damage to the decorations but where, as here, the obligation is merely to keep the structure and exterior of the house in repair, the covenant will only come into operation where there has been damage to the structure and exterior which requires to be made good."***

Lawton LJ said 821G:

"The standard of repair may depend on whether the house is in a South Wales valley or in Grosvenor Square; [I interpolate I am not sure with great respect whether I would wish to be associated with that antithesis] but wherever it is, the landlord need not do anything until there exists a condition which calls for repair. As a matter of the ordinary use of English that which requires repair is in a condition worse than it was at some earlier time. This usage of English is, in my judgment, the explanation for the many decisions on the extent of a landlord's or tenant's obligation under covenants to keep houses in repair. ***Broadly stated, they come to this: a tenant must take the house as he finds it; neither a landlord nor a tenant is bound to provide the other with a better house than there was to start with; but, because almost all repair work requires some degree of renewal, problems of degree arise whether after the repair there is a house which is different from that which was let."***

Dillon LJ was a member of the court directly considering section 4 of the Act in McNerny v Lambeth Borough Council [1989] 19 EG 77. The case concerned the effects of condensation in a flat which caused the plaintiff to suffer from colds and other ailments. The claim failed. Although the single pane windows and solid walls were the cause of the condensation, they could not be said to be in disrepair. Ralph Gibson LJ said at 83M: "But the important point is that that statutory protection for those in occupation of **defective premises is geared to the landlord's obligation to repair the premises. It goes no wider than the repair covenant."**

The approach I favour in this case, on somewhat wider grounds, is I think supported by the reasoning of Chadwick LJ, with whom Tuckey LJ and Sir Murray Stuart-Smith agreed in Lee v Leeds City Council [2002] 1 WLR 1488. **That was a case in which the interiors of two houses suffered from a design fault giving rise to condensation and mould.** *There, as Mr Hartley was this morning at pains to insist, the works that would have been required to make the defects good would have been very substantial indeed. At 1515D and following Chadwick LJ said this:*

"80. But an affirmative answer to the question whether there is to be implied a power to enter in order to carry out remedial works required to remedy defects which pose a danger to health does not lead to the conclusion that failure to carry out such works (if not works of repair) may give rise to liability under section 4(1) of the 1972 Act. As I have said, the first question in this context is whether the works that would be required in order to remedy the inherent defects in design – which are the cause of the excessive condensation and mould – are within the expression 'any description of maintenance or repair of the premises'. ***If they are not, the fact that the landlord may be entitled to enter the premises in order to carry out those works does not give rise to a deemed obligation to the tenant "for that description of maintenance or repair" under section 4(4); so that the defect is not a 'relevant defect' for the purposes of section 4(3); and there is no duty under section 4(1) to take care to see that persons are reasonably safe from injury***

caused by that defect.

"81. In my view the first respondent local authorities are correct in their submission that the first question should be answered in the negative. The works required to remedy the design defects are not works of repair – giving to 'repair' the meaning which it should bear in this context. It is not contended that the works are works of maintenance. It is, I think, significant that Parliament, when enacting section 4 of the 1972 Act in substantially the terms proposed in the draft bill annexed to the Law Commission report Civil Liability of Vendors and Lessors for Defective Premises (1970) (Law Com No 40), chose to link the duty of care imposed by section 4(1) to the landlord's failure to carry out an obligation "for the maintenance or repair" of the premises: see section 4(3). That is the framework within which the statutory hypothesis in section 4(4) must operate. ***Parliament did not, as it might have done, link the duty of care to a failure to remedy defects in any more general sense.*** *The obligation to 'repair' has a well recognised meaning in the law of landlord and tenant; and, as the cases show,* ***<u>it does not</u>*** *<u>arise unless the object in respect of which it is imposed is out of repair</u>. If the defect which has caused the injury in respect of which a claim is made under section 4(1) of the 1972 Act is not a defect arising from of repair, it cannot be a 'relevant defect' for the purposes of that section. As Ralph Gibson LJ put it, in the McAuley case [1992] QB 134, 145: 'There is, I think, no warrant for a wide construction of the words of section 4. They apply to all landlords, and not merely to local authorities, and can operate so as to impose a substantial burden upon a landlord in respect of premises under the immediate control of the tenant and in respect of which the landlord has assumed no contractual obligation.'"*

Key Points:-

a) For the duty to arise under section 4, the landlord must have actual or constructive notice of the defect.

b) The landlord owes an obligation for repair and maintenance of the premises in the lease.

c) The obligation to repair, will not be interpreted wider than the covenant to repair – this means that there is no duty to improve the state of the premises.

d) The duty does not arise unless the object in respect of which it is imposed is out of repair.

e) If a defect does fall within section 4, the claimant can make a claim for damages for the cost of the rectification work, as well as any injury claim.

Claims under Section 1 – The Duty to Construct Dwellings Properly

Section 1 of the 1972 Act sets out:-

1 Duty to build dwellings properly.

(1) A person taking on work for or in connection with the provision of a dwelling (whether the dwelling is provided by the erection or by the conversion or enlargement of a building) owes a duty—
(a) if the dwelling is provided to the order of any person, to that person; and
(b) without prejudice to paragraph (a) above, to every person who acquires an interest (whether legal or equitable) in the dwelling;to see that the work which he takes on is done in a workmanlike or, as the case may be, professional manner, with proper materials and so that as regards that work the dwelling will be fit for habitation when completed.

(2) A person who takes on any such work for another on terms that he is to do it in accordance with instructions given by or on behalf of that other shall, to the extent to which he does it properly in accordance with those instructions, be treated for the purposes of this section as discharging the duty imposed on him by subsection (1) above except where he owes a duty to that other to warn him of any

defects in the instructions and fails to discharge that duty.

(3) A person shall not be treated for the purposes of subsection (2) above as having given instructions for the doing of work merely because he has agreed to the work being done in a specified manner, with specified materials or to a specified design.

(4) A person who—
(a) in the course of a business which consists of or includes providing or arranging for the provision of dwellings or installations in dwellings; or
(b) in the exercise of a power of making such provision or arrangements conferred by or by virtue of any enactment; arranges for another to take on work for or in connection with the provision of a dwelling shall be treated for the purposes of this section as included among the persons who have taken on the work.

(5) Any cause of action in respect of a breach of the duty imposed by this section shall be deemed, for the purposes of the M1Limitation Act 1939, the M2Law Reform (Limitation of Actions, &c.) Act 1954 and the M3Limitation Act 1963, to have accrued at the time when the dwelling was completed, but if after that time a person who has done work for or in connection with the provision of the dwelling does further work to rectify the work he has already done, any such cause of action in respect of that further work shall be deemed for those purposes to have accrued at the time when the further work was finished.

Key Points:-

a) Section 1 requires the landlord to ensure the work which he takes on is done in a workmanlike or, as the case may be, professional manner, with proper materials and so that as regards that work the dwelling will be fit for habitation when completed.

b) The duty is owed by builders, surveyors or other persons undertaking work in connection with a dwelling house.

c) The duty will apply to anyone having a legal or equitable interest in the property.

d) Damages can include damages for economic loss and for the loss of enjoyment of the premises.

e) The cause of action accrues when the work is complete. Limitation will be 6 years, unless there is a personal injury element to the claim.

Section 1 liability works by the Judge making a consideration whether the property as a whole is unfit. For example in *Harrison and Others v (1) Shepherd Homes Ltd (2) National House Building Council (3) NHBC Building Control Services Ltd [2011] EWHC 1811 (TCC),* the Court held that the provision regarding fitness for habitation under Section 1 of the Defective Premises Act 1972 was a measure of the standard required in performance of the relevant duty.

Section 2 of the Act sets out:-

2 Cases excluded from the remedy under section are:

(1)Where—
(a) in connection with the provision of a dwelling or its first sale or letting for habitation any rights in respect of defects in the state of the dwelling are conferred by an approved scheme to which this section applies on a person having or acquiring an interest in the dwelling; *and*
(b) it is stated in a document of a type approved for the purposes of this section that the requirements as to design or construction imposed by or under the scheme have, or appear to have, been substantially complied with in relation to the dwelling; no action shall be brought by any person having or acquiring an interest in the dwelling for breach of the duty imposed by section 1 above in relation to the dwelling.

CHAPTER FOUR
HOUSING DISREPAIR: TAKING INSTRUCTIONS, PRE-ACTION PROTOCOLS, EXPERT EVIDENCE AND DISCLOSURE

Housing disrepair cases can be complex in law. The starting point is to have a form to take instructions and ask the correct questions at the outset of the case. Here is an example of one (this is not designed to be a full questionnaire for any claim but may be a starting point to investigations):-

CLIENT'S DETAILS:

Client's Title and Full Name:	
Client's Full Address:	
Contact Numbers:	
Email Address:	
Date of Birth:	
NI Number:	
Occupation:	

BASIC DETAILS:

Name and Address of Landlord:	
Tenancy Start Date:	
Nature of Property: House or Flat? If flat, what floor? How many floors does the block have? What type of block is it – i.e. new one or old council block? How many bedrooms, bathrooms, living rooms, sitting rooms?	
Whose Name is on the Tenancy Agreement? How many people live in the property? What ages/relationship are these to the client?	
Was property furnished or unfurnished when client moved in? What was condition of the property when you moved in? (If good before, it may be a good claim) If always been this way, don't take on. If increased, take further details.	

How much is the rent per month? Please ask client how often it is paid and by what method – i.e. direct debit, BACS, Card etc.?	
Is the rent up to date? If not, please ask the client how much they owe/how much is in arrears and whether the landlord has threatened to take any action for these arrears. Obtain copies of any correspondence**	

DISREPAIR QUESTIONS:

What is wrong with the property? How does the client know that it is in disrepair? **<u>Please make a list of disrepairs</u>** Examples: - Penetrating water - Presence of damp/mould - Colder in one part of house than another - Peeling wallpaper/discoloured wallpaper - Sewage backwash	

- Missing or dangerous fixtures such as stair banister, uneven floorboards, nails protruding, holes in floorboards - Vermin problems?	
Is there evidence of flaking paint-work, wallpaper or plaster (signs of damp)?	
Is there evidence of black mould or other mould inside the property?	
Is there evidence of increased condensation in parts of the inside of the property?	

POTENTIAL CAUSES OF DISREPAIR:

What is the disrepair? * If it is not clear from above, ask client what the cause of the disrepair is. Ask whether they have any of the following in the property. E.g. missing slates from roof, leading to water coming in, leading to damp etc. - Missing slates from room? - Blocked guttering	

- Missing chimney pots/lids - Eroded pointing to the walls - Boiler not working - Water not working - Draught windows/defective window seals - Defective boiler - Minor leaks	
Are there any other causes for the client's problems? E.g. if the client is complaining of damp – do they ever dry laundry indoors? Do they open windows and vents regularly or at least after having a bath or shower?	

REPAIRS TO DATE:

What if any repairs have been carried out to date? - Please ask client to list these. Need type of repair and date done. - If repairs are ongoing – check whether scaffolding is currently erected at client's house	
Has the Defendant ever offered to re-house the Claimant? If so, what was the offer? Did the Claimant take it? If not, why not?	

Has the client attempted to do anything about the situation? E.g. do they wipe down mould? Use disinfectant? Paint over it? How long did that work for before it came back?	

REPORTING THE DEFECT:

When did the client first become aware of each of the defects? When was it first reported to the landlord? - How was it reported? Obtain exact way in which it was reported, what was done, etc. - All communications / copy of all correspondence if client still has it.	
Has the client ever told a representative of the Defendant about the Defect, e.g. a boiler repair man?	
Has the Landlord ever visited / carried out an inspection of the property since the defects arose? If so:	

- When did they visit? - How often do they visit/inspect?	

EFFECTS OF DEFECT:

How have the defects affected client's use of home? - Have they stopped the client using any of the rooms? - Have they affected any contents? E.g. clothes/items that have been stored ruined because of damp	
Has anyone suffered any physical injury because of the defect? - What Personal Injury? - Has anyone suffered asthma or increased asthma, chest infections, bronchitis or arthritis as a result of the defect? - If so, have they ever visited a GP, walk in centre or sought any other medical treatment? (Get Details if so)	

TIMELINE OF CLAIM:

Date that client first noticed issues:	
List of Issues – where and what are the problems?	
Date that client first reported issues:	
Date that action was first taken in relation to issues	
Result of action – did it fix issues?	
If didn't fix issue, date client reported issue again	
Number of times reported	
Number of failed attempts to fix	
Current Situation	

The Pre-Action Protocol

Claimant lawyers should be aware of the protocol. This must be followed at the outset of a claim and sets out as follows:--

1 INTRODUCTION

1.1 This Protocol applies to residential property situated in England

and Wales. It relates to claims by tenants and others in respect of housing disrepair. Before using the Protocol tenants should ensure that their landlord is aware of the disrepair. The Protocol is intended for those cases where, despite the landlord's knowledge of the disrepair, the matter remains unresolved.

1.2 This Protocol describes the conduct the court will normally expect prospective parties in a housing disrepair claim to follow prior to the start of proceedings. It is intended to encourage the exchange of information between parties at an early stage and to provide a clear framework within which parties in a housing disrepair claim can attempt to achieve an early and appropriate resolution of the issues.

1.3 If a claim proceeds to litigation, the court will expect all parties to have complied with the Protocol as far as possible. The court has power to order parties who have unreasonably failed to comply with the Protocol to pay costs or to be subject to other sanctions.

2 AIMS

2.1 The aims of this Protocol are to–
(a) avoid unnecessary litigation;
(b) promote the speedy and appropriate carrying out of any repairs which are the landlord's responsibility;
(c) ensure that tenants receive any compensation to which they are entitled as speedily as possible;
(d) promote good pre-litigation practice, including the early exchange of information and to give guidance about the instruction of experts; and
(e) keep the costs of resolving disputes down.

3.1 A disrepair claim is a civil claim arising from the condition of residential premises and may include a related personal injury claim (see 3.5 below). Although most claims are brought by a tenant against their landlord, this Protocol is not limited to such claims. It covers claims by any person with a disrepair claim including tenants, lessees and members of the tenant's family. The use of the word "tenant" in this Protocol is intended to cover all such people.

3.2 The types of claim which this Protocol is intended to cover include those brought under ***Section 11 of the Landlord and Tenant Act 1985, Section 4 of the Defective Premises Act 1972, common law nuisance and negligence, and those brought under the express terms of a tenancy agreement or lease.*** *It does not cover claims brought under Section 82 of the Environmental Protection Act 1990 (which are heard in the Magistrates' Court).*

3.3 This Protocol does not cover disrepair claims which originate as counterclaims or set-offs in other proceedings i.e. where the tenant is seeking to have the compensation due for disrepair set against money claimed by the landlord (typically in a possession claim for rent arrears). In such cases the landlord and tenant will still be expected to act reasonably in exchanging information and trying to settle the case at an early stage.

3.4 The Protocol should be followed in all cases, whatever the value of the damages claim.

3.5 Housing disrepair claims may contain a personal injury element. ***If the personal injury claim requires expert evidence other than a General Practitioner's letter, the Personal Injury Protocol should be followed for that element of the disrepair claim. If the personal injury claim is of a minor nature and will only be evidenced by a General Practitioner's letter, it is not necessary to follow the Personal Injury Protocol. If the disrepair claim is urgent, it would be reasonable to pursue separate disrepair and personal injury claims, which could then be case managed together or consolidated at a later date.***

4 ALTERNATIVE DISPUTE RESOLUTION

4.1 The parties should consider whether some form of alternative dispute resolution procedure would be more suitable than litigation

and if so, try to agree which form to use. Both the landlord and the tenant may be required by the court to provide evidence that alternative means of resolving their dispute were considered.
The courts take the view that litigation should be a last resort, and that claims should not be issued while a settlement is still actively being explored. Parties should be aware that the court will take into account the extent of the parties' compliance with this Protocol when making orders about who should pay costs.

4.2 Options for resolving a dispute include the following—
(a) mediation: information about mediation can be found at http://www.civilmediation.org/contact.php
(b) for council tenants—
Local authority repairs, complaints and/or arbitration procedures.
The Right to Repair Scheme. The scheme is only suitable for small, urgent repairs of less than £250 in value. Information and leaflets about the scheme in England can be obtained from the Department for Communities and Local Government, Eland House, Bressenden Place, London SW1E 5DU. Tel: 0303 444 0000 (https://www.gov.uk/repair-council-property).

Information about the scheme in Wales can be obtained from the National Assembly for Wales, Cardiff Bay, Cardiff, CF99 1NA. Tel. 0845 010 5500 http://www.assemblywales.org/index.htm. In England, the Housing Ombudsman Service 81 Aldwych London WC2B 4HN Tel 0300 111 3000 http://www.housing-ombudsman.org.uk/. In Wales the Public Services Ombudsman for Wales. Tel. 0300 790 0203 http://www.ombudsman-wales.org.uk.
(c) for housing association tenants and for tenants of qualifying private landlords—
Any complaints procedure operated by the landlord. In England, the Housing Ombudsman Service, 81 Aldwych, London. WC2B 4HN Tel: 0300 111 3000 http://www.housing-ombudsman.org.uk. In Wales, the National Assembly for Wales, Cardiff Bay, Cardiff, CF99 1NA. Tel. 0845 010 5500 http://www.assemblywales.org/index.htm. Information about repair rights generally is available free of charge from the following web pages:

http://england.shelter.org.uk/get_advice/repairs_and_bad_conditions and http://www.communitylegaladvice.org.uk/en/legalhelp/leaflet04_1.jsp. The former Department for Transport, Local Government and the Regions issued Good Practice Guidance on Housing Disrepair Legal Obligations in January 2002. Copies of the Guidance (ISBN 185112523X) can be obtained from Communities and Local Government Publications, PO Box 236, Wetherby LS23 7NB. Tel: 0300 123 1124. Fax:0300 123 1125. Textphone: 0870 1207 405. E-mail: product@communities.gsi.gov.uk. (free to download from the Communities and Local Government website at http://www.communities.gov.uk/publications/housing/deliveringhousingadaptations2).

A summary, Housing Research Summary No. 154, is available free on the Communities and Local Government website at the following link http://www.communities.gov.uk/archived/publications/housing/housingdisrepair. The Communities and Local Government website http://www.communities.gov.uk is a general source of information for landlords and tenants.

5 TENANT'S LETTER OF CLAIM

5.1 It is recognised that disrepair cases can range from straightforward to highly complex, and that it is not always possible to obtain detailed information at an early stage. In order to avoid unnecessary delay and to ensure that notice of the claim is given to the landlord at the earliest possible opportunity, particularly where repairs are urgent, it may be appropriate for the tenant to send a letter notifying the landlord of the claim before a detailed Letter of Claim is sent.

5.2 The tenant should send to the landlord a Letter of Claim at the earliest reasonable opportunity. A specimen Letter of Claim is at Annex A. The letter may be suitably adapted as appropriate. The Letter of Claim should contain the following details–
(a) the tenant's name, the address of the property, the tenant's

address if different, the tenant's telephone number and when access is available;
(b) details of the defects, including any defects outstanding, in the form of a schedule, if appropriate (See Annex C for a specimen schedule which can be used to inform the landlord of the disrepair);
(c) history of the defects, including any attempts to rectify them;
(d) details of any notification previously given to the landlord of the need for repair or information as to why the tenant believes that the landlord has knowledge of the need for repair;
(e) the effect of the defects on the tenant (including any personal injury claim by the tenant);
(f) the identities of all other persons who plan to make a personal injury claim and brief details of their personal injury claims;
(g) the details of any special damages (see the form at Annex D);
(h) the proposed expert(see paragraph 7);
(i) the proposed letter of instruction to the expert (See Annex B); and
(j) relevant documents disclosed by the tenant.

5.3 The Letter of Claim should also request disclosure from the landlord of all documents relevant to the disrepair including—
(a) a copy of the tenancy agreement including the tenancy conditions;
(b) the tenancy file;
(c) any documents relating to notice of disrepair given, including copies of any notes of meetings and oral discussions;
(d) any inspection reports or documents relating to works required to the property; and
(e) any computerised records.

5.4 Documents relating to rent arrears or other tenancy issues will not normally be relevant. Nothing in the Protocol restricts the right of the tenant to look personally at their file or to request a copy of the whole file. Neither is the landlord prevented from sending to the tenant a copy of the whole file, should the landlord wish.

5.5 A copy of the Protocol should be sent to the landlord if the tenant has reason to believe that the landlord will not have access to the Protocol e.g. because the landlord is an individual or small organisation. If in doubt, a copy should be sent.

6 Landlord's Response

6.1 Where a landlord is not an individual, a person should be designated to act as a point of contact for the tenant (and their solicitor, if one is involved). The designated person's name and contact details should be sent to the tenant and their solicitor as soon as possible after the landlord receives the Letter of Claim from the tenant

6.2 The landlord should normally reply to the Letter of Claim within 20 working days of receipt. Receipt is deemed to have taken place two days after the date of the letter. The landlord's response should include at least the following—
(a) copies of all relevant records or documents requested by the tenant; and
(b) a response to the tenant's proposals for instructing an expert including—
i. whether or not the proposed single joint expert is agreed;
ii. whether the letter of instruction is agreed;
iii. if the single joint expert is agreed but with separate instructions, a copy of the letter of instruction; and
iv. if the appointment of a single joint expert is not agreed, whether the landlord agrees to a joint inspection.

6.3 The landlord must also provide a response dealing with the issues set out below, as appropriate. This can be provided either within the response to the Letter of Claim or within 20 working days of receipt of the report of the single joint expert or receipt of the experts' agreed schedule following a joint inspection—
(a) whether liability is admitted and, if so, in respect of which defects;
(b) if liability is disputed in respect of some or all of the defects,

the reasons for this;
(c) any point which the landlord wishes to make regarding lack of notice of the disrepair or any difficulty in gaining access;
(d) a full schedule of intended works, including anticipated start and completion dates and a timetable for the works;
(e) any offer of compensation; and
(f) any offer in respect of costs.

6.4 Failure to respond within 20 working days of receipt of the Letter of Claim or at all, is a breach of the Protocol (see paragraph 1.3) and the tenant is then free to issue proceedings.

6.5 The Letter of Claim and the landlord's response are not intended to have the same status as a statement of case in court proceedings. Matters may come to light subsequently which mean that the case of one or both parties may be presented differently in court proceedings. Parties should not seek to take advantage of such discrepancies provided that there was no intention to mislead.

*7 **Experts***

General

7.1
(a) Parties are reminded that the Civil Procedure Rules provide that expert evidence should be restricted to that which is necessary and that the court's permission is required to use an expert's report. The court may limit the amount of experts' fees and expenses recoverable from another party.
(b) When instructing an expert, the parties must have regard to CPR Practice Direction 35 and the Guidance for the Instruction of Experts in Civil Claims 2014 at http:www.judiciary.gov.uk.
(c) In some cases it might not be necessary to instruct an expert to provide evidence of disrepair, for example, if the only issue relates to the level of any damages claimed. It may be advisable for tenants to take photographs or video footage of any defects before and after works.

(d) The expert should be instructed to report on all items of disrepair which the landlord ought reasonably to know about, or which the expert ought reasonably to report on. The expert should be asked to provide a schedule of works, an estimate of the costs of repair, and to list any urgent works.

Single Joint Expert

7.2

(a) If the landlord does not raise an objection to the proposed expert or letter of instruction within 20 working days of receipt of the Letter of Claim, the expert should be instructed as a single joint expert, using the tenant's proposed letter of instruction.(See Annex B for a specimen letter of instruction to an expert.)

(b) Alternatively, if the parties cannot agree joint instructions, the landlord and tenant should send their own separate instructions to the single joint expert. If sending separate instructions, the landlord should send the tenant a copy of the landlord's letter of instruction with their response to the Letter of Claim.

Joint Inspection

7.3

(a) If it is not possible to reach agreement to instruct a single joint expert, even with separate instructions, the parties should attempt to arrange a joint inspection, meaning an inspection by different experts instructed by each party to take place at the same time. If the landlord wishes their own expert to attend a joint inspection, they should inform both the tenant's expert and the tenant's solicitor.

(b) Should a case come before the court, it will be for the court to decide whether the parties have acted reasonably in instructing separate experts and whether the costs of more than one expert should be recoverable.

Time Limits

7.4

(a) Whether a single joint expert or a joint inspection is used, the

property should be inspected within 20 working days of the date that the landlord responds to the tenant's Letter of Claim.
(b) If a single joint expert is instructed, a copy of the expert's report should be sent to both the landlord and the tenant within 10 working days of the inspection. Either party can ask relevant questions of the expert who should send the answers to both parties.
(c) If there is a joint inspection, the experts should produce an agreed schedule of works detailing–
i. the defects and required works which are agreed and a timetable for the agreed works
ii. the areas of disagreement and the reasons for disagreement.
(d) The agreed schedule should be sent to both the landlord and the tenant within 10 working days of the joint inspection.

Urgent Cases

7.5 The Protocol does not prevent a tenant from instructing an expert at an earlier stage if this is considered necessary for reasons of urgency. Appropriate cases may include–
(a) where the tenant reasonably considers that there is a significant risk to health and safety;
(b) where the tenant is seeking an interim injunction;
(c) where it is necessary to preserve evidence.
Access

7.6 Tenants must allow the landlord reasonable access for inspection and repair in accordance with the tenancy agreement. The landlord should give reasonable notice of the need for access, except in the case of an emergency. The landlord must give access to common parts as appropriate, for example, for the inspection of a shared heating system. If the tenant is no longer in occupation of the premises the landlord should take all reasonable steps to give access to the tenant for the purpose of an inspection.

Expert's fees

7.7
(a) Experts' terms of appointment should be agreed at the outset, including the basis of charging and time for delivery of the report.
(b) If a single joint expert is instructed, each party will pay one half of the cost of the report. *If a joint inspection is carried out, each party will pay the full cost of the report from their own expert.*
7.8 Information about independent experts can be obtained from—
(a) The Chartered Institute of Environmental Health, Chadwick Court, 15 Hatfields, London. SE1 8DJ Tel: 020 7928 6006 http://www.cieh.org/about_us.html. Ask for a copy of the Consultants and Trainers Directory;
(b) The Expert Witness Directory http://www.sweetandmaxwell.co.uk/our-businesses/directories.aspx
(c) The Royal Institution of Chartered Surveyors, 12 Great George Street, Parliament Square, London. SW1P 3AD, Tel: 024 7686 8555 http://www.ricsfirms.com/ Ask for a copy of the relevant regional directory.

Taking stock

8 Where the procedure set out in this Protocol has not resolved the dispute between the landlord and the tenant, they should undertake a review of their respective positions to see if proceedings can be avoided and, at the least, to narrow the issues between them.

Time limits

9
(a) The time scales given in the Protocol are long stops and every attempt should be made to comply with the Protocol as soon as possible. If parties are able to comply earlier than the time scales provided, they should do so.
(b) Time limits in the Protocol may be changed by agreement. However, it should always be borne in mind that the court will expect an explanation as to why the Protocol has not been followed or has been varied and breaches of the Protocol may lead to costs or other orders being made by the court.

Limitation period

10
(a) There are statutory time limits for starting proceedings ('the limitation period'). If a tenant starts a claim after the limitation period applicable to that type of claim has expired, the landlord will be entitled to use that as a defence to the claim. In cases where the limitation period is about to expire, the tenant should ask the landlord to agree not to rely on a limitation defence, so that the parties can comply with the Protocol.
(b) If proceedings have to be started before the parties have complied with the Protocol, they should apply to the court for an order to stay (i.e. suspend) the proceedings until the steps under the Protocol have been completed.

Costs

11 If the tenant's claim is settled without litigation on terms which justify bringing it, the landlord will pay the tenant's reasonable costs. The Statement of Costs Form N260 available on the HMCTS website can be used to inform the landlord of the costs of the claim.

Example Letters

ANNEX A - LETTER OF CLAIM

(a) Letter from solicitor

To Landlord

Dear Sirs,

RE: TENANT'S NAME AND ADDRESS OF PROPERTY

We are instructed by your above named tenant. (Where the tenant *is legally aided or a party to a conditional fee agreement entered into before 1 April 2013 insert a sentence stating how their case is being funded.) We are using the Housing Disrepair Protocol. We enclose a copy of the Protocol for your information.**

Repairs

Your tenant complains of the following defects at the property (set out nature of defects).

*We enclose a schedule which sets out the disrepair in each room.**

The history of the disrepair is as follows: (set out history of defects)

You received notice of the defects as follows: (list details of notice relied on).

The defects at the property are causing (set out the effects of the disrepair on the client and their family, including any personal injury element. Specify if there will be any additional claimants).

Please arrange to inspect the property as soon as possible. Access will be available on the following dates and times: (list dates and times as appropriate)

Please confirm whether you intend to carry out repairs at this stage or whether you wish to wait until the property has been inspected by the expert(s) as set out below.

If you intend to carry out repairs at this stage, please set out a full schedule of intended works including anticipated start and completion dates and a timetable for the works.

Disclosure

Please also provide within 20 working days of this letter the following:

All relevant records or documents including:
(i) a copy of the tenancy agreement including the tenancy conditions
(ii) the tenancy file
(iii) documents relating to notice of disrepair given, including copies of any notes of meetings and oral discussions
(iv) inspection reports or documents relating to works required to the property.
(iv) computerised records

We enclose a signed authority from our client for you to release this information to us.

We also enclose copies of the following relevant documents from our client:

Expert

If agreement is not reached about the carrying out of repairs within 20 working days of this letter, we propose that the parties agree to jointly instruct a single joint expert (insert expert's name and address) to carry out an inspection of the property and provide a report. We enclose a copy of their CV, plus a draft letter of instruction.

Please let us know if you agree to his/her appointment. If you object, please let us know your reasons within 20 working days.

If you do not object to the expert being instructed as a single joint expert, but wish to provide your own instructions, you should send those directly to (insert expert's name) within 20 working days of this letter. Please send us a copy of your letter of instruction. If you do not agree to a single joint expert, we will instruct (insert expert's name) to inspect the property in any event. In those circumstances, if you wish to instruct your expert to attend at the same time, please let us and (insert expert's name) know within 20 working days.

Claim

We take the view that you are in breach of your repairing obligations. Please provide us with your proposals for compensation. (Alternatively, set out suggestions for general damages i.e. £x for x years).

*Our client also requires compensation for special damages, and we attach a schedule of the special damages claimed.**

Yours faithfully,

ANNEX B: LETTER OF INSTRUCTION TO EXPERT

(a) Letter from solicitor

Dear

RE: TENANT'S NAME AND ADDRESS OF PROPERTY

We act for the above named in connection with a housing disrepair claim at the above property. We are using the Housing Disrepair Protocol.

*We enclose a copy of the Protocol for your information.**

*Please carry out an inspection of the above property by (date)** and provide a report covering the following points:*

(a) whether you agree that the defects are as claimed;
(b) whether any of the defects is structural;
(c) the cause of the defect(s);
(d) the age, character and prospective life of the property;
(e) a schedule of works;
(f) an estimate of the costs of repair.

Access will be available on the following dates and times: (list dates and times as appropriate).

*You are instructed as a single joint expert / The landlord is (landlord's name and details) / The landlord will be providing you with their own instructions direct / The landlord will contact you to confirm that their expert will attend at the same time as you to carry out a joint inspection.**
Please provide the report within 14 days of the inspection. Please contact us immediately if there are any works which require an interim injunction.

If the case proceeds to court, the report may be used in evidence. Please ensure that the report complies with Civil Procedure Rules Practice Direction 35.3 and the Guidance for the Instruction of Experts in Civil Claims 2014 at http:www.judiciary.gov.uk.

If you do not have a copy please let us know.

Insert details as to cost and payment

Yours sincerely,

** Delete as appropriate*
*** The date to be inserted should be 20 working days from the date of the letter*

The Pre Action Protocol also contains a very helpful room by room schedule of repairs which can be downloaded from https://www.justice.gov.uk/courts/procedure-rules/civil/pdf/protocols/housing-disrepair-protocol-annex-c-schedule-of-disrepair.pdf and a special damages form, which can be downloaded from https://www.justice.gov.uk/courts/procedure-rules/civil/pdf/protocols/housing-disrepair-protocol-annex-d-special-damages-Form.pdf to be sent with the letter of claim.

CHAPTER FIVE
HOUSING DISREPAIR: CAUSATION, REMEDIES, QUANTUM AND COSTS IMPLICATIONS

<u>Remedies - Order for Works</u>

Section 17 of the LTA 1985 says:-

> *17 Specific performance of landlord's repairing obligations.*
>
> *(1) In proceedings in which a tenant of a dwelling alleges a breach on the part of his landlord of a repairing covenant relating to any part of the premises in which the dwelling is comprised, the court may order specific performance of the covenant whether or not the breach relates to a part of the premises let to the tenant and notwithstanding any equitable rule restricting the scope of the remedy, whether on the basis of a lack of mutuality or otherwise.*
>
> *(2) In this section—*
> *(a) "tenant" includes a statutory tenant,*
> *(b) in relation to a statutory tenant the reference to the premises let to him is to the premises of which he is a statutory tenant,*
> *(c) "landlord", in relation to a tenant, includes any person against whom the tenant has a right to enforce a repairing covenant, and*
> *(d) "repairing covenant" means a covenant to repair, maintain, renew, construct or replace any property.*

This will likely apply to any repairs that have not been completed by the date of Trial. The duty to repair under section 4 of the DPA 1972, can also be enforced by an order for repairs. If the landlord defaults, an order can be obtained for Contempt of Court.

The Court can also make a declaration at Trial that:-

1. The landlord is in breach of the repairing obligation;

2. The tenant is entitled to do the necessary repair works of repair in default; and/or

3. The work can be financed from deductions of future rent.

The Claimant Lawyer would also need to make sure that any settlement order includes an obligation to carry out the works in a specific time period, and the terms under which the works are to be done.

General Damages

The claimant in a housing disrepair case can claim damages to put the tenant back in the positon had the breach not occurred. The principles were explored in JULIE WALLACE V MANCHESTER CITY COUNCIL (1998) as follows:-

> *"In 1989 the plaintiff, Mrs. Wallace, went, with her two small children, to live in property belonging to Manchester City Council at 62 Abergele Road, Fallowfield, Manchester. On 22nd August 1994 she was given a written tenancy agreement. This stated that her rent was £45.26 per week. In July 1995 the weekly rent was increased to £47.40 per week. Thus at all times since 1989 Mrs. Wallace has been a secure weekly tenant and, as such, subject to the provisions of Part IV Housing Act 1985.*
>
> *2. There were two relevant obligations on the part of the City Council implied into such tenancy. The first was an obligation to keep in repair the structure and exterior of the dwelling house and the installations inside the dwelling house as prescribed by and on the terms set out in s.11 Landlord & Tenant Act 1985. The second subjected the City Council to a duty to take such care as is reasonable in all the circumstances to see that all persons who might reasonably be affected by defects in the state of the premises are reasonably safe from personal injury or from damage to their property caused by a relevant defect as defined in s.4 Defective Premises Act*

1972.

3. On 8th March 1996 the property was inspected by Hiltons, building surveyors, on the instructions of the solicitors acting on behalf of Mrs Wallace. They reported "Discussions with the Tenant revealed the following matters, namely:

- The external wall below the Living Room window has collapsed some 2 to 3 months ago.
- The windows are generally rotten. That to the Living Room is in a particularly poor condition. As a consequence the Living Room is always extremely cold despite the heating being on continuously.
- A remedial Damp Proof Course has been installed on 3 occasions to the Living Room, the last being in August 1995. There is a mould build up behind the Tenants side board to the side party wall.
- Plaster and skirtings are deficient/falling off/loose to the Bedrooms.
- The Tenant has suffered from constant infestation from rats which enter the property from the drains at the rear of the house. The Council have attended on 3 occasions to lay preventative substances. The operative has however stated that the rats are eating the laid substances and that it may be possible for the infestation to re- occur.
- The rainwater pipe to the front elevation leaks badly during periods of heavy rain.

We understand that these deficiencies have existed over a number of years with the Tenant having complained to the Council but gained no effective response."

4. On 27th August 1996 the City Council's building surveyor inspected the property and prepared a schedule of works. In December 1996 repairs were carried out by way of replacing a window and repointing a wall, but, the City Council contend, they were unable to obtain access so as to carry out further works.

5. On 24th January 1997 Mrs Wallace and her two children commenced proceedings in the Manchester County Court claiming that the City Council was in breach of each of the two obligations to which I have referred. They alleged that Mrs Wallace had complained to the City Council about the want of repair on many occasions since 1992. This was a material allegation for the liability of the landlord under both statutory provisions depends on having notice of the want of repair or of the relevant defect as the case may be. They alleged that each of them had suffered distress, anxiety and inconvenience in consequence. They claimed specific performance of the repairing obligations and damages for diminution in the value of Mrs Wallace's rent, for inconvenience in respect of all of them and for the ill- health suffered by the children.

6. On 2nd July 1997 the action came before His Honour Judge Hardy. He heard evidence from Mrs Wallace, her surveyor, Mr Williams and the building surveyor for the City Council, Mr Claringbold. In his judgment, given on the same day, the judge recorded that to a large extent the disrepair had not been disputed and that he was not invited to make detailed findings or orders in that respect. He found that there had been breaches of both the obligations relied on with the consequence that Mrs Wallace and her children were entitled to damages. He assessed the damages to be paid to each child at £2,000 and the special damages, for damage to curtains, carpets and furniture, in the amount claimed, namely £780. The judge continued: "That leaves two heads of damages which are pursued. **The one is the award of general damages to the plaintiff and it is said to the children for the general inconvenience and diminution in value it is said to the value of the property to them.** *I am invited to consider dealing with an award partially on the basis of an order for general damages and partially under what can be termed perhaps a subhead of general damages, not exactly special damages but to calculate it on some basis which assess them as the diminution in the value of the tenancy to them in relation to the amount of rent paid.*

I am hesitant about doing it that way. I know that other Judges in concurrent jurisdictions have done so and it may be a useful exercise to consider it in that way as part of the way in which one arrives at a figure of general damages. I do not regard it myself as a separate head of damages from general damages but a part of them and if Judges wish to use that method in arriving at their total figure they, of course, so far as I am concerned [are] perfectly at liberty to do so. Personally speaking, I do not find it a useful approach.

Added to which it presents what, to my mind, is a considerable difficulty. If this lady had been paying the whole of the rent out of her own pocket the argument might have additional force but for the majority of the time she was not paying any of the rent out of her own pocket but it was paid with housing benefit. For a period of about two years she was contributing towards it at £18 per week.

Damages are intended, in so far as money can, to place people in the same position as if the [wrong] had not occurred. It seems to me that it is a breach of that general principle to award to a person a sum calculated by relation to a figure which has been paid to a defendant when it has not been paid by the plaintiff. I can foresee all sorts of problems in doing it this way. It would in effect reward the plaintiff by a figure which she had not suffered as a cash sum. **It seems to me far better, rather than use that approach, to award her a sum which would be adequate compensation in my view for the damage she has suffered which is the distress, inconvenience and disruption to her lifestyle and that of her children by the defects complained of. That seems to me a far better way of approaching it."**

7. After referring to such assistance as may be obtained from considering awards in other cases the judge awarded damages not only for the discomfort and distress already encountered but also for the further discomfort when the remedial works were being carried out in the total sum of £3,500 for all heads of general damage. Subsequently the repairs set out in a schedule agreed between the parties' respective surveyors were carried out. Mrs. Wallace has remained the tenant of the dwelling house.

10. I shall, in due course, consider the cases to which Counsel for Mrs Wallace referred us in this **connection. But the agreed starting point is the basic principle that the purpose of an award of damages is, so far as possible by an award of money, to place the innocent party in the position he would have been in if he had not suffered the wrong, whether breach of contract or tort, of which he complains.** *We were referred to passages to this effect in Johnson v Agnew [1979] 2 WLR 487, 499 in the context of damages in lieu of or in addition to a decree of specific performance and Dodd Properties (Kent) Ltd v Canterbury City Council [1980] 1 AER 928 in relation to repairs to property necessitated by the commission of a tort.* **Thus in the context of the breach of an obligation to repair property let to another the purpose of the award of damages is, so far as an award of money can do so, to place the tenant in the position he or she would have been in if the obligation to repair had been properly performed.**

11. Counsel for Mrs Wallace relied on six decisions of this court. They were McCoy v Clark [1982] 13 HLR 87; Calabar Properties Ltd v Stitcher [1984] 1 WLR 287; Sturolson v Mauroux [1988] 20 HLR 332; Chiodi v De Marney [1988] 21 HLR 6; Ezekiel v McDade [1995] 2 EGLR 107 and Brent LBC v Carmel Murphy [1995] 28 HLR 203. I will refer to them in that order.

12. In McCoy v Clark [1982] 13 HLR 87 the plaintiff landlord sued for arrears of rent and the defendant tenant counterclaimed for damages in respect of disrepair to the roof over his flat. The judge accepted the submission that the disrepair to the roof meant that the flat was worth substantially less to the tenant than it would have been if the roof had been in order. **He assessed the damages on the basis of a proportion of the rent payable under the tenancy, namely 10% for the first 113 weeks and 20% for the succeeding 68 weeks.** *The tenant appealed on the grounds that such an award was too low. The Court of Appeal accepted that* *submission. Sir David Cairns, with whom Stephenson LJ agreed, considered that the figure found by the judge was too low. At page 94 he said*

"I am satisfied, not only that the judge did find, but that it is the right finding, that the main cause of this dampness and the main cause of the flat being worth less to the defendant than it otherwise would have been was the plaintiff's breach in relation to the roof.

On that basis, was the compensation that was awarded to the defendant for it adequate? In my view it was not. It is all very well to say that the defendant was not spending a great deal of the day in the flat and that he was using it mainly as a sleeping place. If he had the flat as a sleeping place and was willing to pay £9 a week for the flat for that purpose, then he is entitled to a flat which is comfortable for that purpose, and if it is substantially reduced in the degree of comfort, then I think what he ought to recover is something proportional to that reduction."

Counsel for Mrs. Wallace accepted that the case shows no more than that the assessment of damages by reference to a proportion of the rent was a method acceptable to the court. *In particular there was no separate claim or award for discomfort and inconvenience. For my part I do not think that the case is authority which supports either side; the point with which we are concerned was not in issue. But insofar as it favours either of them it supports the City Council because Sir David Cairns evidently considered that a proportion of the rent might be used as a measure of the degree of comfort foregone. To that extent comfort and convenience and reduction in value of the flat were regarded as opposite sides of the same coin.*

13. In Calabar Properties Ltd v Stitcher [1984] 1 WLR 287 the plaintiff landlord sued for arrears of rent. The tenant, the assignee of the residue of a term of 99 years at a rent of £100 per annum, counterclaimed for damages for breach by the landlord of his covenant to keep the block of which the tenant's flat was part in good and substantial repair. The Official Referee decided that the dampness in the flat of which the tenant complained was caused by water penetration attributable to the landlord's persistent failure to comply with his covenant. He also concluded that the tenant was justified in moving to alternative accommodation while the repairs were being carried out. **The judge awarded damages of £4,606 for the cost of redecoration of the flat, which he used as a measure of diminution in value, and £3,000 for discomfort and loss of enjoyment.** *He refused to award any damages to cover the running costs of the flat when unoccupied nor for any consequential loss of use based on either the capital value of the flat or its rack rental value. The tenant appealed contending that the judge should have awarded damages under those heads also.*

14. Counsel for Mrs Wallace relied on this case as showing that a tenant might obtain an award of general damages for both the cost of redecoration, as representing the diminution in value of the lease, and for discomfort and inconvenience to the tenant occasioned by the landlord's failure to repair. This is true; but, first, there was no appeal against the awards of damages the judge had made only against those he had refused and, second, Stephenson LJ (at page 290H), with whom May LJ agreed, regarded the award of damages for the cost of redecoration as an award of special, not general, damages.

15. The issues the court was required to decide do not arise in this case. Nevertheless two members of the court made observations in relation to the question of the recoverability of damages for consequential loss of use which are germane. The tenant had contended that she was entitled to a further award for such damages based on the diminution in the capital or rack rental value of the flat occasioned by the landlord's failure to repair in accordance with his covenant. The contention was based on a statement of Bankes LJ in Hewitt v Rowlands (1924) 93 LJKB 1080, 1082 that **"...the measure of damages is the difference in value to the tenant of the premises, from the date of the notice to repair down to the date of the assessment of damages, between the premises in**

their present condition and their value, if the landlord on receipt of the tenant's notice had fulfilled the obligations of the covenant."

Rejecting the submission for the tenant Stephenson LJ, at page 293 said:

"The second objection is that to submit that what the defendant has lost by the plaintiffs' breach of covenant is the consequent diminution in the value of the flat as a marketable asset is to ask the court to take a wholly unreal view of the facts. The reality of the defendant's loss is the temporary loss of the home where she would have lived with her husband permanently if the plaintiffs had performed their covenant. She cannot increase her loss by deciding not to return after the covenant has been performed, and she does not seek to do so. But she can claim, as it seems to me, to be put in as good a position as she would have been if the plaintiffs had performed their covenant, as least as early as they had notice that the main structure was out of repair instead of years later. If she had bought the lease as a speculation intending to assign it, to the knowledge of the plaintiffs, the alleged diminution of rental (or capital) value might be the true measure of her damage. But she did not; she bought it for a home, not a saleable asset, and it would be deplorable if the court were bound to leave the real world for the complicated underworld of expert evidence on comparable properties and value, on the fictitious assumption that what the flat would have fetched had anything to do with its value to her or her husband. I do not think we are bound by the authority of Hewitt v Rowlands, 93 L.J.K.B. 1080, or any other decision to do something so absurd."

Later, at page 295, he added in relation to the decision in Hewett v Rowlands:

"What is plain is that, in laying down the measure of damage, the court cannot have had the capital or rental value of the cottage as a marketable asset in mind, because a statutory tenancy is not market-

able, and the court was considering the position of a statutory tenant who was still living in the cottage and would lose his interest in the cottage if he ceased to live there. What the difference in value to the plaintiff of the statutory tenancy of the cottage repaired and unrepaired may have been was not an easy matter for the registrar to assess, but I suspect he would not have gone far wrong if he had equated it with what the plaintiff might have to spend on performing the landlord's covenant (assuming the landlord would not perform it himself) and substantial general damages for inconvenience and discomfort for the period from notice to the landlord till assessment or performance of the covenant by the plaintiff. In my judgment, there is nothing in that case which supports Mr. Ralls' claim to any such additional sum as he claims for diminution in value to the tenant or which would disable the judge from measuring that diminution and the defendant's damage by the amounts he has awarded as general and special damages - and by the reasonably incurred costs of alternative accommodation, an item which was never in issue in Hewitt v Rowlands, 93 L.J.K.B. 1080, because the tenant never left the cottage he occupied.

17. At page 297 Griffiths LJ stated, "**The object of awarding damages against a landlord for breach of his covenant to repair is not to punish the landlord but, so far as money can, to restore the tenant to the position he would have been in had there been no breach.** *This object will not be achieved by applying one set of rules to all cases regardless of the particular circumstances of the case. The facts of each case must be looked at carefully to see what damage the tenant has suffered and how he may be fairly compensated by a monetary award.*

In this case on the findings of the judge the plaintiff landlords, after notice of the defect, neglected their obligation to repair for such a length of time that the flat eventually became uninhabitable. It was also clear that unless ordered to do so by an order of the court, the plaintiffs had no intention of carrying out the repairs. In these circumstances the defendant had two options that were reasonably open to her: either of selling the flat and moving elsewhere, or alternatively of moving into temporary accommodation and bringing an action against the plaintiffs to force them to carry out the repairs, and then returning to the flat after the repairs were done. If the defendant had chosen the first option then the measure of damages would indeed have been the difference in the price she received for the flat in its damaged condition and that which it would have fetched in the open market if the plaintiffs had observed their repairing covenant. If however the defendant did not wish to sell the flat but to continue to live in it after the plaintiffs had carried out the necessary structural repairs it was wholly artificial to award her damages on the basis of loss in market value, because once the plaintiffs had carried out the repairs and any consequential redecoration of the interior was completed there would be no loss in market value. The defendant should be awarded the cost to which she was put in taking alternative accommodation, the cost of redecorating, and some award for all the unpleasantness of living in the flat as it deteriorated until it became uninhabitable. These three heads of damage will, so far as money can, compensate the defendant for the plaintiffs' breach"

Later, at page 299, of the statement of Bankes LJ which I have quoted, he commented:

"Whatever Bankes L.J. meant by "the difference in value to the tenant," the one thing he cannot have meant in the circumstances of that case was the diminution in the market value of the tenancy, for it was a statutory tenancy which the tenant could not sell, and thus it had no market value. In my view the difference in value to the tenant must vary according to the circumstances of the case. ***If the tenant is in occupation during the period of breach he is entitled to be compensated for the discomfort and inconvenience occasioned by the breach and I suspect that that is what Bankes L. J. had in mind when he used the phrase "the difference in value to the tenant"*** *Hewitt v Rowlands, 93 L.J.K.B. 1080, 1082, for which the judge in this case awarded £3,000.* ***If the tenant has rented the property to let it and the landlord is aware of this, then "the difference in value to the tenant" may be measured by***

his loss of rent if he cannot let it because of the landlord's breach. *If the tenant is driven out of occupation by the breach and forced to sell the property then "the difference in value to the tenant" may be measured by the difference between the selling price and the price he would have obtained if the landlord had observed his repairing covenant. But each case depends upon its own circumstances and Hewitt v Rowlands should not be regarded as an authority for the proposition that it is in every case necessary to obtain valuation evidence.*

In my view there was no need for any valuation evidence in this case. I repeat that damages in a case such as this should include the cost of the redecoration, a sum to compensate for the discomfort, loss of enjoyment and health involved in living in the damp and deteriorating flat and any reasonable sum spent on providing alternative accommodation after the flat became uninhabitable."
May LJ agreed with both judgments.

18. In my view, whilst not constituting any binding authority on the point we have to decide, the dicta which I have quoted do not support the contention of Mrs Wallace. As in McCoy v Clark so in Calabar Properties Ltd v Stitcher it was recognised that where the tenant wishes to remain in occupation of the property the diminution in value occasioned by the landlord's failure to repair for which he is entitled to be compensated is the personal discomfort and inconvenience he has experienced as a result of the want of repair. It is evident that the court considered, and I agree, that the assessment of the amount of money necessary to compensate the tenant was a matter for the judge and not for expert evidence.

19. In Sturolson & Co v Mauroux [1988] 20 HLR 332 the landlord sought to recover arrears of rent. The defendant, a statutory tenant, counterclaimed for damages for breach by the landlord of his covenant to repair and to provide certain services. By the time of the hearing the necessary repairs had been carried out. The judge awarded damages under two heads, namely £1,345 for diminution in value calculated as a percentage of the rent payable and the balance for discomfort, inconvenience and injury to health. The landlord appealed contending that the rent officer had taken the want of repair into account when assessing the fair rent so that the award of damages gave rise to double recovery. He also contended that the tenant had failed to mitigate his loss. Both points were rejected. Counsel for Mrs Wallace relies on this case as indicating the approval of this court to awards of damages on both the bases for which he contends. It is true that the court did not disapprove of those two awards, no party suggested that it should. But I do not accept that the decision is any support for the proposition that a judge must award damages under both those heads.

20. In Chiodi v De Marney [1988] 21 HLR 6 a statutory tenant occupying a flat at a registered fair rent of £8 per week withheld the rent and was sued for possession. He counter-claimed for damages for breach of the implied covenant on the part of the landlord to repair. The judge awarded him damages under three heads of which only one is relevant for present purposes. That head was for inconvenience and distress calculated on the basis of £30 per week for hree and a half years. The landlord appealed contending that, as the rent was only £8 per week, the award was too high. The appeal was rejected. The case shows that a weekly sum, even in excess of the rent payable for the premises, is a permissible way to calculate the monetary compensation to be awarded for distress and inconvenience. It is no support at all for the proposition that a court is bound to assess damages under the heads of both inconvenience and diminution in value.

21. In Ezekiel v McDade [1995] 2 EGLR 107 the court awarded damages for a negligent valuation on the basis of diminution in value and discomfort and distress. It was suggested that it was in some way supportive of the submissions made on behalf of Mrs Wallace. I can only say that I disagree.

22. Finally counsel for Mrs Wallace relied on Brent LBC v Carmel Murphy [1995] 28 HLR 203. In that case the defendant, a secure tenant, when sued for arrears of rent counterclaimed for damages for breach of the landlord's obligation to repair. The judge awarded her general damages for diminution of the value of the tenancy calculated by reference to a reduction in the rent payable and general damages by reference to an annual sum. Counsel for Mrs Wallace relies on the circumstance that there was there an award on both the bases for which he contends and leave to appeal was refused. One of the grounds of appeal was that the awards of damages were excessive. Roch LJ said that there was no indication that the awards were wrong in principle or excessive in amount. Plainly this case also indicates that damages may be awarded on both bases. Likewise, in my view, it is no support for the proposition that the court is bound to do so.

23. I have dealt at length with the cases relied on by counsel for Mrs Wallace because of the importance of the point raised on this appeal to District and County Court Judges throughout England and Wales. I can express my conclusions more shortly in the form of a series of propositions. First, the question in all cases of damages for breach of an obligation to repair is what sum will, so far as money can, place the tenant in the position he would have been in if the obligation to repair had been duly performed by the landlord. Second, the answer to that question inevitably involves a comparison of the property as it was for the period when the landlord was in breach of his obligation with what it would have been if the obligation had been performed. Third, for periods when the tenant remained in occupation of the property notwithstanding the breach of the obligation to repair the loss to him requiring compensation is the loss of comfort and convenience which results from living in a property which was not in the state of repair it ought to have been if the landlord had performed his obligation. (McCoy v Clark; Calabar Properties Ltd v Stitcher and Chiodi v De Marney) Fourth, if the tenant does not remain in occupation but, being entitled to do so, is forced by the landlord's failure to repair to sell or sublet the property he may recover for the diminution of the price or recoverable rent occasioned by the landlord's failure to

perform his covenant to repair. (Calabar Properties Ltd v Stitcher)

24. Obviously the tenant cannot claim damages in accordance with the third proposition for periods occurring after the sale or sub-lease referred to in the fourth. To that extent, as shown in Calabar Properties Ltd v Stitcher, those two heads are mutually exclusive. This case is concerned with the proper application of the third proposition, not the fourth.

Thus the question to be answered is what sum is required to compensate the tenant for the distress and inconvenience experienced because of the landlord's failure to perform his obligation to repair. Such sum may be ascertained in a number of different ways, including but not limited to a notional reduction in the rent. Some judges may prefer to use that method alone (McCoy v Clark), some may prefer a global award for discomfort and inconvenience (Calabar Properties Ltd v Stitcher and Chiodi v De Marney) and others may prefer a mixture of the two (Sturolson v Mauroux and Brent LBC v Carmel Murphy). But, in my judgment, they are not bound to assess damages separately under heads of both diminution in value and discomfort because in cases within the third proposition those heads are alternative ways of expressing the same concept.

25. It follows that in my judgment Judge Hardy was right when he said that diminution in the value of the property in relation to the amount of rent paid is not a separate head of damage. In the light of the submissions made to us I would make some general observations on the problems of assessing damages in this field. First, I would agree with the observations of Stephenson and Griffiths LJJ in Calabar Properties Ltd v Stitcher that expert valuation evidence is not of assistance when assessing the damages in accordance with my third proposition. The question is the monetary value of the discomfort and inconvenience suffered by the tenant. That is a matter for the judge. As Kennedy LJ observed in the course of argument there is no market in out of repair council houses on which expert evidence could be either admissible or helpful. Second, a judge who seeks to assess the monetary compensation to be awarded for discomfort and inconvenience on a global basis would be well advised to cross-check his prospective award by reference to the rent payable for the period equivalent to the duration of the landlord's breach of covenant. By this means the judge may avoid over- or under-assessments through failure to give proper consideration to the period of the landlord's breach of obligation or the nature of the property".

Key Points from the authorities cited in the case of Wallace:-

a) An award can be made for distress, inconvenience and disruption, as well as loss of enjoyment arising from the dwelling.

b) The claimant will normally recover damages from the date of disrepair to the date the repairs are undertaken.

c) If the repairs are still not undertaken by the date of Trial, damages will be recovered to the date of Trial.

d) The object of damages, is to place the claimant back into the position, had the obligations been properly performed.

e) Damages could be a global award for discomfort and inconvenience.

f) Damages could be a notional reduction in rent, or could be a reduced rental value and award for discomfort/inconvenience.

g) Damages are not designed to be punitive.

h) If the tenant is in occupation during the period of breach he is entitled to be compensated for the discomfort and inconvenience occasioned by the breach.

i) If the tenant has rented the property to let it and the landlord is aware of this, then "the difference in value to the tenant" may be measured by his loss of rent if he cannot let it because of the landlord's breach.

j) A weekly sum, even in excess of the rent payable for the premises, is a permissible way to calculate the monetary compensation to be awarded for distress and inconvenience. It is no support at all for the proposition that a court is bound to assess damages under the heads of both inconvenience and diminution in value.

k) The Judge is not bound to use either calculation method, but a reduction in value of rent may be the most straight forward method of undertaking the calculation.

l) Damages will need to be updated for inflation.

m) Any damages post 1 April 2013, will obtain the 10% uplift as in *Castle v Simmons.*

Quantum of General Damages

There are many reported cases in Kemp and Kemp, but here are just a selection of the main Court of Appeal cases for disrepair. Figures must be updated for RPI.

In PAUL ARNOLD V GREENWICH LONDON BOROUGH COUNCIL (1997), Greenwich London Borough Council was found to be in breach of an express covenant to maintain a dwelling in good condition and had to pay **£8,750 in general damages to a tenant who had endured mould, smell and noise for six years.**

In JONATHAN EARLE v SOTOS CHARALAMBOUS (2006) –

> *"The appeal concerned the method by which damages should be assessed for an admitted breach of repairing covenant by a lessor. The "central issue", as it was described by the appellant, was* ***whether the judge had been correct in assessing the damages to***

take account of rental value, or whether he should have treated himself as restricted by reference to a "tariff" based on previous awards. *The appellant failed on that issue.*

However he succeeded in a more limited attack on the reasoning of the judge in arriving at his assessment for period 1. For that period, he had rejected the lessee's own methodology, but had arrived at "a global figure", without any precise indication of how it was calculated. We held that his award of £20,000 for that period was not supported by adequate reasoning, and was excessive when viewed in relation to the lessee's own claim.

In order to arrive at a substituted figure of damages, we suggested that the starting point should be the judge's finding of rental value, taken with the lessee's own assessment of the percentage degree of impact. *But we added:*

"However, this simple approach needs some qualification. First, while the judge did not accept that there had been a "step- change" in September 2000, some account must in my view be taken of the lessee's perception that things got worse over time, although I bear in mind also that his calculation included a separate figure for "stress and depression". Secondly, for the reasons I have given (and subject to any contrary submissions), some allowance must be made for the time reasonably needed for repairs in any event. ***In the absence of clear findings on either point, a fair approach might simply be to disregard the initial period up to September 2000 for the purpose of the calculation. This would lead to an award for Period 1 equivalent to 50% of the rental value from September 2000 until December 2002 (27 x £1,000 x 50%), giving a total of £13,500."***

TANYA GRAND v PARAM GILL (2011):-

Ms Grand was formerly Mr Gill's tenant of the second floor (top) flat in a three-storey building at 9B Marlborough Parade, Uxbridge Road, Hillingdon, Middlesex. The ground floor consists of commer-

cial premises and there is another flat on the first floor. She originally had a 12-month assured shorthold tenancy under a tenancy agreement dated 21 November 2004; and following the expiration of the original term, the tenancy continued as an assured periodic tenancy. Ms Grand occupied the flat with her daughter Alison, who was 17 at the time of the trial in May 2009. **The flat comprises two bedrooms and a living room, kitchen, bathroom and lavatory. Mr Gill has a long leasehold interest in the flat but no interest in the remainder of the building. Ms Grand made claims against him under several heads.** *They included claims for harassment under the Protection from Harassment Act 1997 and for personal injury, which both failed and with which this appeal is not concerned.* **She also advanced a claim for breach of Mr Gill's covenant for quiet enjoyment, which succeeded and for which she recovered damages of £350 that formed part of the total damages awarded of £5,600.** *There is no need to refer further to that head of claim. I turn to the issues before the judge with which the appeal is concerned.*

33. I now turn to assess the damages. ... In assessing the damage and disrepair I have regard to and follow the approach approved by the Court of Appeal in Wallace v. Manchester City Council [1998] 3 EGLR 38. **I have regard to the rent that was payable. I have regard to the discomfort and inconvenience.** *I approach this case on the basis that it is right to take both aspects into account, so that I approach it as a mixture of the two.* **I take into account the rent in assessing damages for distress, discomfort and inconvenience. The most important damage, so to speak, was the consequence of the damp and mould, but it seems to me that on the evidence it is not possible to say that more than a ten per cent contribution to the damp and mould was made by the lack of adequate heating coupled with the small matters of the ingress of water into the living room from the guttering and the broken window which was repaired after a year or so. On the basis of full liability for that part of the claim I would have considered that damage in the region of £2,000 per annum would be appropriate, that is to say, £6,000 in all, but since I** find the defendant is responsible only to the extent of ten per cent, the award under this head is only £600.

34. A much bigger award arises in relation to the boiler problem. The 207 days, amounting to approximately 30 weeks, during which there was no heating should, in my judgment, be dealt with by an award of damages of £1,750. The remainder of the first three years of the tenancy, during which there was inadequate heating coming from the boiler, should be compensated at the rate of £1,200 per annum. In setting that figure I have regard to the decision mentioned in Islington v. Spence in July 2001, Legal Action 26 in Clerkenwell County Court, referred to by [counsel for Mr Gill] in paragraph 43 of his skeleton argument, in which the court awarded £1,100 per annum for the period during which, after heating had been restored, the radiators failed to heat up properly, and I add a little to take account of [inaudible] since 2001, so that produces a figure of £3,600 for three years, less approximately £700 for 30 weeks covered by the award of damages for there being no heating at all, so producing a total under this head of £2,900. This means that damages for disrepair will be awarded at a total figure of £5,250

.

35. As to the damages for breach of the covenant for quiet enjoyment, in the circumstances it seems to me it is a relatively modest sum. It will be appropriate to cover the distress and inconvenience suffered by [Ms Grand] and I fix that at £350. So the total damages will be £5,600.'

Key Points:-

a) A good starting point will be the Judges' finding of rental value – i.e. the rent due under the tenancy for the period of disrepair.

b) The second stage will be to look at what percentage of the property was affected and also what level of impact the disrepair had on the property – i.e. the loss of use and enjoyment of the tenant. For example in *Gill v Grand* – lack of heating that caused damp and mould was 10% – the Court said:- " *it cannot be said that more than a ten per cent contribution to the damp and mould was made by the lack of adequate heating coupled with the small matters of the ingress of water into the living room from the guttering and the broken window which was repaired after a year or so".* It will be important the claimant's statement deals in detail with this aspect. However, no heating for 30 weeks, amounted to an award of £1,750.

c) The Court can order a separate award for discomfort and inconvenience – for example in *Grand v Gill* it was £350. This will depend upon the level of disrepair. Kemp will likely assist with specific awards in specific circumstances. Again, the claimant's statement will need to deal in detail with this aspect.

d) An easy way of calculating an award may be – if there for example the annual rent in a property is £5,000 – there is damp, mould or condensation that affects 50% of the property. The period of disrepair is 12 months. An award for a combination of reduced rental value and discomfort / inconvenience might be £2,500.

e) Awards will vary accordingly to the level of disrepair.

f) It is worth remembering on a case run under a Conditional Fee Agreement, the client may have to pay an ATE premium, if there is no LEI/BTE cover of between £600-900, as the damages are likely to be limited to an award for reduced value of rent, distress or inconvenience. Monetary compensation is not likely to be recovered for actual repairs as the landlord will be required to carry these out. If the average rent on a property is say £500 per month or £6,000 per annum and, for example, only approximately 20% of the property was affected by the disrepair i.e. one room, the award could be in the region of £1,200 for a 12 month repair claim for perhaps damp or mould affecting one room on the reduced rental analysis (plus the landlord being ordered to carry out the repairs).

However, if the disrepair was limited to 6 months – the award for loss of enjoyment, distress or inconvenience could be as little as £600. This may not cover the ATE premium. Claimant lawyers may be wary about accepting a minor case of say minor damp / condensation / mould (with no personal injury claim), where the period of disrepair is less than 6 months.

This may not apply for example to a claim where there is no heating at all, where the awards could be higher.

Special Damages

A claim for housing disrepair may include a claim for special damages. Examples of such are, extra heating costs, loss of earnings, cleaning costs, costs of redecoration, costs of items damaged in the property and costs of alternative accommodation, should it have been required. Claimant lawyers should ask tenants to list all items caused by the disrepair. However, it should be noted that the claimant/tenant will need to provide evidence in support of these heads of loss.

CHAPTER SIX
HOUSING DISREPAIR: CASES IN POINT, LITIGATION ISSUES, LIMITATION AND COSTS

Limitation Periods

For any claims for personal injury, limitation will be three years from the date of injury, or date of knowledge, where the injury was caused by disrepair in the property. This limitation period applies to the entirety of the claim, including the housing disrepair element, should a personal injury claim be involved.

When the premises are in disrepair, for claims under the Housing Act 1985, limitation will run for six years from when the landlord had knowledge of the problem and had failed to remedy it. Claims should always be started within that period. However, as the cause of action starts afresh every day that the Landlord fails to remedy the defect it is possible to bring a claim for a disrepair that started more than six years ago but the claim can only go back for the last six years prior to issue of proceedings in those cases.

For claims under section 1 of the Defective Premises Act 1972, limitation will run from the date of completion of the construction works.

For damage claims under section 4 of the DPA 1972, limitation will run from the date of damage.

If a disrepair claim involves an injury and other heads of damage – if all the claims are brought within one action, section 11 of the Limitation Act 1980 will apply a three year limitation period to all causes of action, including inconvenience and distress. Section 11 sets out:-

Special time limit for actions in respect of personal injuries.

(1) This section applies to any action for damages for negligence, nuisance or breach of duty (whether the duty exists by virtue of a contract or of provision made by or under a statute or independently of any contract or any such provision) where the damages claimed by the plaintiff for the negligence, nuisance or breach of duty consist of or include damages in respect of personal injuries to the plaintiff or any other person.

[F1(1A) This section does not apply to any action brought for damages under section 3 of the Protection from Harassment Act 1997.]

(2) None of the time limits given in the preceding provisions of this Act shall apply to an action to which this section applies.

(3) An action to which this section applies shall not be brought after the expiration of the period applicable in accordance with subsection (4) or (5) below.

(4) Except where subsection (5) below applies, the period applicable is three years from—
(a) the date on which the cause of action accrued; or
(b) the date of knowledge (if later) of the person injured.

(5) If the person injured dies before the expiration of the period mentioned in subsection (4) above, the period applicable as respects the cause of action surviving for the benefit of his estate by virtue of section 1 of the M1Law Reform (Miscellaneous Provisions) Act 1934 shall be three years from—
(a) the date of death; or
(b) the date of the personal representative's knowledge; whichever is the later.

(6) For the purposes of this section "personal representative" includes any person who is or has been a personal representative of the deceased, including an executor who has not proved the will (whether or not he has renounced probate) but not anyone

> *appointed only as a special personal representative in relation to settled land; and regard shall be had to any knowledge acquired by any such person while a personal representative or previously.*
>
> *(7) If there is more than one personal representative, and their dates of knowledge are different, subsection (5)(b) above shall be read as referring to the earliest of those dates.*

The following section of the CPR will be relevant to housing disrepair claims:-

> *(4) In a claim which includes a claim by a tenant of residential premises against a landlord where the tenant is seeking an order requiring the landlord to carry out repairs or other work to the premises, the claimant must also state in the claim form –*
> *(a) whether the estimated costs of those repairs or other work is –*
> *(i) not more than £1,000; or*
> *(ii) more than £1,000; and*
> *(b) whether the value of any other claim for damages is –*
> *(i) not more than £1,000; or*
> *(ii) more than £1,000.*

Expert Evidence

Expert evidence will normally be from a chartered civil surveyor and will need to be attached to the Particulars of Claim, together with a copy of the tenancy agreement. CPR 35 rules relating to Expert Evidence will be applicable.

Costs

QOCS will apply to any claim for personal injury, but probably not for the housing disrepair element.

Where there is no personal injury claim, QOCS will not apply and the Claimant will need ATE insurance to cover the risk of having to pay the Defendant's costs. If there is no BTE/LEI cover available, it is a good idea to have an ATE policy in place that will cover all the defendant's costs (for the non PI elements of the claim), and unrecovered disbursements.

Housing Disrepair Case Examples – Condensation and Damp/Mould

There are many examples of complaints for housing disrepair. It could relate to drains or sewers, sanitation, gas or electricity services. One of the most common examples is condensation or damp. This was dealt with in the case of *Quick v Taff Ely BC* where the Court confirmed that to succeed the claimant has to show disrepair of the structure or exterior of the property, causing the dampness or condensation. This was further examined in the case of *RUTH LEE V LEEDS CITY COUNCIL : RATCLIFFE & ORS V SANDWELL METROPOLITAN BOROUGH COUNCIL (2002)*, where the Court said:-

> *10. The other members of the Court (Lord Justice Lawton and Lord Justice Neill) agreed with Lord Justice Dillon. Lord Justice Lawton said this, at page 823B–C:*
>
> *"In my judgment, there must be disrepair before any question arises as to whether it would be reasonable to remedy a design fault when doing the repair. In this case, as the trial judge found, there was no evidence that the single-glazed metal windows were in any different state at the date of the trial from what they had been in when the plaintiff first became a tenant. The same could have been said of the lintels. The judge misdirected himself in finding that these windows required repair."*
>
> *There is no doubt that the Court recognised that the effect of the condensation in the Quick case was that parts of the house (at the least) were unsuitable for occupation. Lord Justice Dillon described the living conditions of the plaintiff and his family as "appalling" - ibid , at page 815B. Lord Justice Lawton referred - at page 821C – to the uncontradicted evidence, accepted by the trial judge, that the sitting room could not be used because of the smell of damp.* ***But it is equally clear that this Court has held, in the Quick case, that***

the exterior and structure is not out of repair by reason only of the fact that as a result of internal condensation - unrelated to water penetration - the dwelling has become unfit for human habitation"

The decision in the Quick case was applied by this Court in Post Office v Aquarius Properties Ltd [1987] 1 All ER 1055 - see at pages 1063h and 1065d . In Stent v Monmouth District Council (1987) 19 HLR 269, at page 285, Sir John Arnold, President, explained the effect of those decisions:

It is no doubt the case that where there is no repair requiring to be done as in such cases as Quick v Taff Ely B.C. [1985] 3 WLR 981 and Post Office v Aquarius Properties Ltd (1985) 276 EG 923 . . . **the repairing covenant on its true construction does not require any design defect to be made good."**

In Southwark London Borough Council v Tanner [2001] 1 AC 1 Lord Hoffmann treated as settled law the proposition that the **repairing covenant to be implied under section 11 of the Landlord and Tenant Act 1985 did not require the landlord to ensure that the premises were fit for the purpose for which they had been let.** *He said this, at page 8B–C:*

It is true that in each tenancy agreement the council agreed to keep the structure in repair . Such an obligation would in any case be implied by section 11 of the Landlord and Tenant Act 1985. But the appellants do not rely upon this covenant and cannot do so. **Keeping in repair means remedying disrepair. The landlord is obliged only to restore the house to its previous good condition. He does not have to make it a better house than it originally was***: see Quick v Taff Ely Borough Council [1986] QB 809."*

"That was the application of that principle which led this Court to hold, in ***the Stent case****, that the landlord was required to eliminate the design defect which had caused a wooden exterior door to rot;*

> *and to hold, in Staves & Staves v Leeds City Council (1990) 23 HLR 107, that the landlord was required to do the works necessary to prevent further damage to internal plaster from condensation and damp.* ***But it is important to keep in mind that, in those cases, there was evidence of damage to the structure (the damaged plaster being treated as part of the structure for that purpose) which gave rise to the need for repair. The cases show that, where there is a need to repair damage to the structure, the due performance of the obligation to repair may require the landlord to remedy the design defect which is the cause of the damage.*** ***<u>They do not support the proposition that the obligation to repair will require the landlord to remedy a design defect which has not been the cause of damage to the structure; notwithstanding that the defect may make the premises unsuitable for occupation or unfit for human habitation".</u>***

The DPA Act 1972 did not assist: -

> *15. The effect of section 4(1) of the 1972 Act is that the duty to take reasonable care to see that persons are kept reasonably safe from personal injury caused by a relevant defect is owed not only to the tenant but also to "all persons who might be expected to be affected by defects in the state of the premises". Persons who might be expected to be affected by defects in the state of the premises will include members of the tenant's family living at the premises. But the duty under section 4(1) arises only where personal injury or damage to property is caused by a "relevant defect". For that purpose - and subject to the provisions in section 4(4) of the Act – a "relevant defect" means a defect in the state of the premises arising from, or continuing because of, an act or omission by the landlord "which constitutes . . . a failure by him to carry out his obligation to the tenant for the maintenance or repair of the premises". It follows - but, again, subject to the provisions in section 4(4) - that where* **the defect which has caused the personal injury or damage to property is a "design defect" which does not constitute in law a "disrepair" for which the landlord is responsible under his repairing obligations (whether express or implied), there is no**

liability under section 4(1) of the 1972 Act.

16. Section 4(4) of the 1972 Act extends liability under section 4(1) to certain defects in the state of the premises which do not arise from the landlord's failure to carry out his repairing obligations. It does so by introducing a statutory hypothesis. A landlord who is not otherwise under an obligation to carry out works of maintenance or repair of a particular description is to be treated - for the purposes of sections 4(1) and (3) but for no other purpose - as if he were under an obligation to carry out works of maintenance and repair of that description if the terms of the tenancy (expressly or impliedly) give him "the right to enter the premises to carry out [that] description of maintenance or repair of the premises". So, for example, a landlord who is not otherwise under an obligation to carry out works of repair and maintenance to the garden but who has a right to enter for the purpose of carrying out such works if he chooses will be treated - for the purposes of section 4(1) and (3) of the Act - as if he were under an obligation to carry out such works; with the consequence that he owes a duty to take reasonable care to see that the tenant or other person who might reasonably be expected to be affected by defects in the state of the garden is reasonably safe from personal injury caused by such a defect arising from lack of repair or maintenance - see McAuley v Bristol City Council [1992] 1 QB 134".

Key Points on Mould, Condensation or Damp Cases:-

a) There is no duty to improve the state of the premises. The landlord only has to keep in repair what is already there.

b) There must be actual damage to the structure or exterior of the property that is causing the damp or mould.

c) Landlords will often argue that it was caused by the tenant by for example use of a tumble dryer without ventilating.

d) The surveyor will need to identify what the defect is that has caused the damp, mould or condensation.

e) The 1985 Act does not require the landlord to ensure the quality of the premises or that they were fit for the purpose they were let.

f) The landlord is not required to remedy a design defect or construction defect that was pre-existing, ***which has not been the cause of damage to the structure.***

g) Where there is a need to repair damage to the structure, the due performance of the obligation to repair may require the landlord to remedy the design defect which is the cause of the damage.

h) Section 4 of the Defective Premises Act 1972 will not assist where there is no actual disrepair for which the landlord is liable.

In *Lee* the Court helpfully summarised the position as follows:-

The LTA 1985:-

> *"68. Nor do I think that an obligation to remedy design defects which lead to excessive condensation and mould can be regarded as a necessary correlative to the tenant's qualified obligations in respect of internal decorative repair. The tenant's obligations must be conditioned by the nature of the premises. If the premises suffer from condensation and mould by reason of some inherent design defect, then the tenant cannot be required to do more by way of decoration than is reasonable in those circumstances. To hold that the landlord is obliged to put the premises into a state in which they are free from condensation so that the tenant can decorate them would be to impose on the landlord an obligation which goes far beyond anything that can properly be regarded as correlative to the tenant's obligations"*

CHAPTER SEVEN
WHAT IS A CAVITY WALL? A TECHNICAL EXPLANATION (UPDATED TO INCLUDE CLAIMS QUESTIONNAIRE)

Cavity walls were first introduced on the wind driven rain exposed west and southwest coastal areas of the UK in the latter part of the nineteenth century, to stop wind driven rain from penetrating through the solid built brick walls of homes. The traditional cavity wall is built with two skins of brickwork, an inner and an outer leaf separated by a void or air gap (the cavity). These two skins are usually tied together with metal wall ties, or brick ties at specific intervals, the latter brick or stone ties can allow penetrating dampness and cold bridging across the cavity that can be further exacerbated by retro fitting insulation.

Cavity walls gradually gained popularity and became the preferred method of construction for external load bearing walls across the whole of the UK because the air gap in the cavity was also found to provide a degree of thermal insulation to properties and, in addition, provided ventilation.

Since the 1970s this insulation quality was further enhanced by the use of lightweight, aerated blocks, rather than bricks, to build the inner leaf of the cavity wall. The primary reason for building cavity walls has always been to keep penetrating wind driven rain out, a principle that has almost been forgotten in our modern world, mostly to our detriment.

In the 1980s the UK Building Regulations introduced requirements for new houses to be built with increasing levels of insulation materials within the cavity. Since this time consecutive governments have encouraged home-owners to insulate their homes in an attempt to save on energy usage and its direct impact upon our environment.

Building Regulations insist upon cavity wall insulation systems in new build properties. These systems must include a clear cavity, as well as the insulation to allow wind driven rainfall to trickle down the internal face of the outer leaf of the cavity to prevent the moisture from contacting the internal leaf.

Successive governments have encouraged the owners of older properties to introduce retrofitted Cavity Wall Insulation within their homes and there have been grants and incentives for homeowners made available under various schemes. Retrofitted cavity wall insulation is full fill and is in direct contact with the wet outer leaf of the cavity wall and this is where serious problems can start to occur.

Retrofit cavity wall insulation systems have been tested and certificated as not allowing the transfer of moisture from one side of the cavity to another. However, it is being reported time and again that wet insulation products are holding water and wreaking havoc within people's homes. Added to this is the fact that these properties are older, may not be maintained to a reasonable standard, and over the years alterations or window replacements will have introduced rubble and debris into the cavity wall acting as a further bridge to penetrating dampness. It therefore becomes clear why we are hearing more and more about cavity wall insulation failure occurring within properties that have had this retrofitted insulation.

CAVITY WALL QUESTIONNAIRE

CLIENT'S DETAILS:

Client's Title and Full Name:	
Client's Full Address:	
Contact Numbers:	
Email Address:	
Date of Birth:	
NI Number:	
Occupation:	

BASIC DETAILS:

Is client a homeowner or tenant? **Client must be homeowner**	
Address for Cavity Wall Insulation Installation	

Date of installation. Did client agree to the installation, or, was it a previous occupant that had it installed?	
Installation Company Please get name, address and reference if there is one	
Date of Contract for Installation	
Does client still have a copy of the installation contract?	
Does client still have a copy of the CIGA guarantee the insulation came with?	

PROPERTY QUESTIONS:

* Explain to client that we will get a surveyor's report, however, need to ask a few question to get a clearer idea of the damage

How does the client know that there is a problem with the Cavity Wall Insulation Installation? Has client ever had surveyor's report re Cavity Wall issues?	
When did they first notice the problem?	

How long after installation did the problem become apparent?	
Is there evidence of damp in the property? If so: (a) When did the client first notice this? (b) How severe is it? (c) Where is it in the property? (d) How big are the patches of mould or damp? (e) Was it noticed after the Cavity wall was installed? If so, how long after?	
Is the dampness of the south facing or west facing wall? *more likely to be evidence of a failed cavity wall, if south or west facing.	
Is there evidence of damp – i.e. any flaking paintwork, wallpaper or plaster in the property? If so, as above – where is it? How big are the patches? How long have they been there?	
Is there evidence of black mould or other mould inside the property? If so, as above – where, how big, how long has it been there? Before or after cavity wall?	

Is there evidence of increased condensation in the property? As above – where, how severe, Before or after cavity wall installation?	
Does client's house seem to have become noticeably colder following the installation of Cavity Wall? If so, when did they notice that? Which parts of the house are the coldest? *Inform client that damp insulation in cavities can actually draw out the heat, rather than keeping it in.* Has client seen any increase in heating costs too since installation?	
How does the client know that the issues aren't something else, other than cavity wall problems?	
i. Is the property exposed to severe levels of wind-driven rain? ii. Is it located in an unsheltered position, i.e. not protection by trees or other buildings iii. Would client said that the external walls are well built? iv. Are there cracks in the brickwork or rendering?	

Inform client that guidance published by the Building Research Establishment says that in these cases, there is an increased risk of rain penetration if a cavity is fully filled with insulation.' Potentially a sign that house wasn't suitable for the insulation, as, the rain can penetrate the outer wall and transfer moisture to internal walls, causing damp	
Did client have any issues with the property, before the cavity wall insulation was fitted? Did client have any damp/mould before insulation was fitted? If so, how severe was it?	
When did the client first realise that issues could be related to faulty cavity wall installation? i.e. - when was the date of knowledge?	
Has the Cavity Wall Insulation caused any external damage to the building? If so, how, where and what?	
Has the Cavity Wall Insulation caused any internal damage to the building? If so, how, where and what?	

POTENTIAL CAUSES – THE INSTALLATION:

What type of Cavity Wall Insulation did the client have?	
Did client get blown mineral fibre wool put into the property? **n.b. - This is the cause of most problems**	
At the time of the install, did client see material? Did client ever see wool being blown into the cavity by a contractor, using a high-pressure jet straight into the wall?	
Was client provided with a copy of a site survey before the installation? Does client remember anyone coming down a site survey, e.g. a surveyor? If so, name and address of surveyor, if they still have it.	
Was there anything during the install that caused client concern? Any problems?	
Following installation: (f) Was the client ever told about a guarantee for the insulation through CIGA? (g) Was the client given a copy of the guarantee?	

(h) Did anyone ever come to sign off the work?	
Has the client ever complained to CIGA or the installer? If so: (i) Do they have copies of correspondence/response? What was the response?	

QUANTUM:

Has the client had an estimate for extraction? If so, how much is it going to cost?	
Has the client had an estimate for repairs – i.e. to fix the damp, wallpaper, paint, etc.? If so, how much?	
Has the client tried and failed to sell the property, due to the problems? If so, outline how	
Does the client have an injury claim arising out of damp or condensation in the property.	

* Enquire whether asthma or any breathing difficulties arisen as a result of living in damp house.	
Has client got any other specific financial losses as a result of the Cavity Wall issues?	

CHAPTER EIGHT
CAVITY WALL CLAIMS: RELEVANT STANDARDS, ESTABLISHING BREACH OF DUTY AND CAUSATION (UPDATED TO INCLUDE CIGA CLAIMS)

Cavity wall insulation is installed as a system. The manufacturer of the insulation is known as a System Designer. It is a requirement of their System Designer Accreditation, issued by the British Board of Agrément, that they approve, train, vet and police the installer that will install their specific system. They will also carry out on-site audits to ensure that the system is being delivered as per the accreditation.

Who are the British Board of Agrément?

The British Board of Agrément was formed on 2nd May 1966 by the Ministry of Public Buildings and Works to assess, test and certify new building products. Today BBA certification is now respected across the building material supply and construction industry and it is a symbol of quality and assurance. It is a vital tool for material manufacturers and System Designers to gain credibility for their products and get them specified as the preferred choice by architects, specifiers, building control, local authorities and insurers.

When the BBA approve a specific system, they issue an Agrément Certificate on that product.

Agrément Certificates enable a manufacturer's new product or system to gain quick recognition in its intended marketplace. Only products and systems that have passed a series of robust assessments – including inspections during the production process, quality management auditing, vigorous testing in a laboratory, and on-site testing, are awarded an Agrément Certificate. Products that have been awarded Agrément Certificates are recognised by local authorities, government departments, architects, specifiers and industry insurers. When a product or system has received a BBA Agrément certificate, the system is audited at regular intervals to ensure the maintenance of quality and standards, the certificate is fully reviewed every three years.

What gives rise to a Cavity Wall Claim?

There are two distinct requirements that would determine whether or not a cavity wall failure could be pursued as a claim:-

a) Breach of Duty;

b) An actual loss to the customer/homeowner.

Breach of Duty

There must be a clear and provable breach of the system design, or proof that the installer failed to properly assess the suitability of the property prior to install, which again, is a breach of the system. If we look at the requirements for a fairly typical Agrément Certificate for Cavity Wall Insulation we see some specific actions that the installer must carry out and some specific details that they must consider when assessing the suitability of a property for cavity wall insulation:

The following are some examples and some explanation of where breaches can and do occur:

> *This Agrément Certificate Product Sheet(1) relates to [NAME] Cavity Wall Insulation, a granulated glass mineral wool fibre material injected in loose form, for use in external masonry cavity walls up to and including 12 m in height, with cavity widths not less than 50 mm, in existing domestic and non-domestic buildings.*
>
> *This system must not be installed in any cavity less than 50mm in width, where it has been would constitute a breach of the system.*
>
> *[NAME]* Cavity Wall Insulation is satisfactory for use as an injected cavity wall insulation and is effective in reducing the thermal trans-

> mittance (U value) of *external cavity walls, with masonry inner and outer leaves (where masonry includes clay and calcium silicate bricks, concrete blocks, natural and reconstituted stone blocks).* The product is for use in existing domestic and non-domestic buildings up to and including 12 m.

If the system has been installed in walls not of masonry construction say in timber framed properties or system built non-standard properties, then this would be a breach of the system

Prior to the installation, an assessment is carried out by a trained assessor, who may also be the installing technician to ascertain the suitability of the property or properties to receive [NAME] Cavity Wall Insulation. *An assessment report is prepared and held at the installer's offices. Particular problems are specifically identified and any reasons for rejection of the work noted. Care should be taken at this stage for the assessor and therapy commissioning the work to identify, and agree in writing as appropriate, any areas of the wall that will not be filled (see section 17.7) and any special requirements for making good (see section 17.5).*

Again, should the assessment report not be completed and held at the installer's office with all relevant customer signatures, then again this is a breach of the system.

The assessment is an important one as the property must have been deemed suitable for insulation, if later it is found that it was not suitable, that maybe it had no Damp Proof Course, that the cavities contained debris and were not clear or that they were less than 50mm in width then the cumulative effects of this breach are quite damming on the installer and impossible to effectively defend.

<u>Loss to the Customer/Homeowner as a direct result of the Breach</u>

The customer must have suffered a loss as a direct result of the defective installation of Cavity Wall Insulation. This could be penetrating dampness as a result of the walls being in a poor condition or dampness caused by a compromised damp proof course. Remember when an installer installs within a house that had pre-existing damp problems, they are effectively declaring the house damp free, as if they acknowledge the damp they should walk away from the job.

Typical damage would be:

- Damage to plaster and decorative finishes
- Damage to timber members through rotting.
- Where a pre-cast concrete home has been installed wrongly and should never have been done, the drilling process will have damaged and weakened the entire structure, and is very difficult to repair, therefore we may have to consider the loss being a diminution in market value or an assumed cost to repair the concrete panels.

<u>Could this liability be covered by the Installers Operational Policy of Insurance?</u>

If we look at standard wordings/definition for public liability insurance, then what is covered is loss or damage to persons or property during the course of the installers business. It then follows, that if the property has been damaged, then, the damage caused could be considered under the installer's public liability insurance. The argument against this from insurers appears to be that there is no cover for defective workmanship. This may be so, however, there is cover for damage and loss as a result of the companies actions or omissions. Some installers carried Professional Indemnity Insurance to cover them in the event that a claim for professional negligence were brought against them, i.e. if they had failed to identify a property was not suitable for Cavity Wall Insulation and installed it anyway leading to loss and damage to the property.

It is worth noting the two very distinct characteristics of both Public liability Insurance and Professional Indemnity Insurance. Public liability cover continues for the period cover was paid for retrospectively, even if the company ceases to exist. Professional indemnity insurance, unless 'run off' by the administrators of a failed company will cease when the premium is not paid and all cover for claims back into time, will cease too.

Guarantees

OFGEM, which stands for Office of Gas and Electricity Markets, is a UK government department and an independent Regulatory Authority. Their principal role when carrying is to protect the interests of gas and electricity customers. Successive Energy Company Obligated schemes created by the Department for Energy, Business & Industrial Strategy and administered by OFGEM require that an appropriate guarantee be placed upon Cavity Wall Insulation Installs, by installers where the install is being funded by any of the big six energy companies under the Energy Company Obligation derived funding schemes. There are numerous guarantee providers that are approved by OFGEM to offer an appropriate guarantee for Cavity Wall Insulation installs, however, by far the most widely used is CIGA, the Cavity Insulation Guarantee Agency.

CIGA have been in the spotlight of the media (and parliament) as failing to properly honour guarantees and failing to acknowledge the true extent of Cavity Wall Insulation failure issues. This, some would say, is not surprising as this 'not for profit' organisation is managed and operated by senior board members of many of the insulation manufacturers and system designers and, as such, is considered by many not to be independent and with a vested interest in protecting its stakeholders. Guarantees can only be invoked when an installer fails to deal with any issues.

Update – CIGA Claims

Consumers do have a right to seek redress through CIGA, if they have an appropriate guarantee in place – https://ciga.co.uk/. The CIGA scheme however does not provide a monetary award, so it is unlikely to be suitable for a CFA claim. However, a client who has a claim that is not suitable as the installer is insolvent / insured could be directed to the scheme as a last resort. More information on CIGA can be found at https://ciga.co.uk/. Legal costs are not recoverable against the scheme. It is also possible that legal costs can be recovered in any arbitration conducted under the CIGA scheme where CIGA refuses to act on the guarantee, for example, where CIGA states that the problem is being caused by the owner's lifestyle. Such arbitration proceedings could potentially be very expensive and it is recommended that specialist counsel is used to assist in such cases. Arbitration proceedings against CIGA are beyond the scope of this book.

CHAPTER NINE
CAVITY WALL CLAIMS: PRE-ACTION PROTOCOLS, DISCLOSURE AND EXPERT EVIDENCE (UPDATED TO INCLUDE EXPERTS DUTIES AND CHOOSING THE CORRECT SURVEYOR)

A key difference between cavity wall claims and housing disrepair claims is that the claimant in a cavity wall claim is the owner of the property, rather than the tenant and thus, his claim does not come within the exception to the £10,000.00 small claims limit that applies to housing disrepair claims.

Accordingly, unless the claim is worth more than £10,000.00, it will usually fall into the Small Claims Track subject to any decision by the court to put the matter into the Fast Track due to complexity. It will be rare, in the authors' opinions, for a cavity wall claim for less than £10,000.00 to be put into the Fast Track. Many claims against builders, for example, relating to badly laid driveways or poorly constructed conservatories etc. are brought in the Small Claims court and cavity wall claims are not materially more complex than these. An exception to this would be if the claim proceeded as a product liability claim against the designer where the designer is located out of the jurisdiction which would be complex enough to engage the Fast Track and possibly even the Multi Track.

Solicitors should remind themselves of the procedure and rules governing the Small Claims Track as they may not be familiar with these from their general practice. CPR Part 27 may be found at https://www.justice.gov.uk/courts/procedure-rules/civil/rules/part27#27.14 and the relevant Practice Direction at https://www.justice.gov.uk/courts/procedure-rules/civil/rules/part27/pd_part27#A

The Practice Direction sets out as follows what evidence should be provided, where relevant to a cavity wall claim. It is worth remembering that, unless ordered otherwise, the normal limit on recovering experts' fees in the Small Claims Track is £750.00.

BUILDING DISPUTES, REPAIRS, GOODS SOLD AND SIMILAR CONTRACTUAL CLAIMS (where the information or documentation is available)

- any written contract;
- photographs;
- any plans;
- a list of works complained of;
- a list of any outstanding works;
- any relevant estimate, invoice or receipt including any relating to repairs to each of the defects;
- invoices for work done or goods supplied;
- estimates for work to be completed;
- a valuation of work done to date.

Whether in the Small Claims Track or in the Fast Track the pre-action protocol for Construction and Engineering Disputes will apply and this can be found at: https://www.justice.gov.uk/courts/procedure-rules/civil/protocol/prot_ced and is set out here:-

1 Introduction

1.1 This Pre-Action Protocol applies to all construction and engineering disputes (including professional negligence claims against architects, engineers and quantity surveyors).

2 Exceptions

2.1 A Claimant shall not be required to comply with this Protocol before commencing proceedings to the extent that the proposed proceedings (i) are for the enforcement of the decision of an adjudic-

ator to whom a dispute has been referred pursuant to section 108 of the Housing Grants, Construction and Regeneration Act 1996 ("the 1996 Act"), (ii) include a claim for interim injunctive relief, (iii) will be the subject of a claim for summary judgment pursuant to Part 24 of the Civil Procedure Rules, or (iv) relate to the same or substantially the same issues as have been the subject of recent adjudication under the 1996 Act, or some other formal alternative dispute resolution procedure.

2.2 A Claimant shall not be required to comply with this Protocol before commencing proceedings if all the parties to the proposed proceedings expressly so agree in writing.

3 Objectives

3.1 The objectives of this Protocol are:

3.1.1 to exchange sufficient information about the proposed proceedings broadly to allow the parties to understand each other's position and make informed decisions about settlement and how to proceed;

3.1.2 to make appropriate attempts to resolve the matter without starting proceedings and, in particular, to consider the use of an appropriate form of ADR in order to do so.

4 Compliance

4.1 If proceedings are commenced, the Court will be able to treat the standards set in this Protocol as the normal reasonable and proportionate approach to pre-action conduct. It is likely to be only in exceptional circumstances, such as a flagrant or very significant disregard for the terms of this Protocol, that the Court will impose cost consequences on a party for non-compliance with this Protocol.

5 Proportionality

5.1 The overriding objective (CPR rule 1.1) applies to the pre-action period. The Protocol must not be used as a tactical device to secure advantage for one party or to generate unnecessary costs. In many cases, including those of modest value, the letter of claim and the response can be simple and the costs of both sides should be kept to a modest level. In all cases, the costs incurred at the Protocol stage should be proportionate to the complexity of the case and the amount of money which is at stake. The Protocol is not intended to impose a requirement on the parties to marshal and disclose all the supporting details and evidence that may ultimately be required if the case proceeds to litigation.

6 Overview of the Protocol

General aim

6.1 The general aim of this Protocol is to ensure that before Court proceedings commence:

6.1.1 the Claimant and the Defendant have provided sufficient information for each party to know the outline nature of the other's case;

6.1.2 each party has had an opportunity to consider the outline of the other's case, and to accept or reject all or any part of the outline case made against him at the earliest possible stage;

6.1.3 there is more pre-action contact between the parties;

6.1.4 better and earlier exchange of information occurs;

6.1.5 there is better pre-action investigation by the parties;

6.1.6 the parties have usually met formally on at least one occasion; and

6.1.7 the parties are in a position where they may be able to settle cases early, fairly and inexpensively without recourse to litigation; and

6.1.8 proceedings will be conducted efficiently if litigation does become necessary.

7 The Letter of Claim

7.1 Prior to commencing proceedings, the Claimant or his solicitor shall send to each proposed Defendant (if appropriate to his registered address) a copy of a letter of claim which shall contain the following information:

7.1.1 the Claimant's full name and address;

7.1.2 the full name and address of each proposed Defendant;

7.1.3 a brief summary of the claim or claims including (a) a list of principal contractual or statutory provisions relied on (b) a summary of the relief claimed including, where applicable, the monetary value of any claim or claims with a proportionate level of breakdown. The extent of the brief summary should be proportionate to the claim. Generally, it is not expected or required that expert reports should be provided but, in cases where they are succinct and central to the claim, they can form a helpful way of explaining the Claimant's position;

7.1.4 the names of any experts already instructed by the Claimant on whose evidence he intends to rely identifying the issues to which that evidence will be directed; and

7.1.5 the Claimant's confirmation as to whether or not it wishes the Protocol Referee Procedure to apply as provided at paragraph 11 below.

8 The Defendant's Response

The Defendant's acknowledgment

8.1 Within 14 calendar days of receipt of the letter of claim, the Defendant should acknowledge its receipt in writing and may give the name and address of his insurer (if any) and shall also confirm whether or not it wishes the Protocol Referee Procedure as provided at paragraph 11 below to apply. If there has been no acknowledgment by or on behalf of the Defendant within 14 days, the Claimant will be entitled to commence proceedings without further compliance with this Protocol.
Objections to the Court's jurisdiction or the named Defendant

8.2 If the Defendant intends to take any objection to all or any part of the Claimant's claim on the grounds that (i) the Court lacks jurisdiction, (ii) the matter should be referred to arbitration, or (iii) the Defendant named in the letter of claim is the wrong Defendant, that objection should be raised by the Defendant within 28 days after receipt of the Letter of Claim. The letter of objection shall specify the parts of the claim to which the objection relates, setting out the grounds relied on, and, where appropriate, shall identify the correct Defendant (if known). Any failure to take such objection shall not prejudice the Defendant's rights to do so in any subsequent proceedings, but the Court may take such failure into account when considering the question of costs.

8.3 Where such notice of objection is given, the Defendant is not required to send a letter of response in accordance with paragraph 8.5 in relation to the claim or those parts of it to which the objection relates (as the case may be).

8.4 If at any stage before the Claimant commences proceedings, the Defendant withdraws his objection, then paragraph 8.5 and the remaining part of this Protocol will apply to the claim or those parts of it to which the objection related as if the letter of claim had been received on the date on which notice of withdrawal of the objection had been given.

The Defendant's Response

8.5 Within 28 days from the date of receipt of the letter of claim, the Defendant shall send a letter of response to the Claimant which shall contain the following information:

8.5.1 A brief and proportionate summary of the Defendant's response to the claim or claims and, if the Defendant intends to make a Counterclaim, a brief summary of the Counterclaim containing the matters set out in paragraph 7.1.3 above;

8.5.2 the names of any experts already instructed on whose evidence it is intended to rely, identifying the issues to which that evidence will be directed;

8.5.3 the names of any third parties the Defendant intends to/is considering submitting to a Pre-action Protocol process.

8.6 If no response is received by the Claimant within the period of 28 days, the Claimant shall be entitled to commence proceedings without further compliance with this Protocol.
Claimant's Response to Counterclaim

8.7 The Claimant shall provide a Response to any Counterclaim within 21 days of the Defendant's Letter of Response. The Response shall contain a brief and proportionate summary of the Claimant's Response to the Counterclaim.

9 Pre-Action Meeting

9.1 Within 21 days after receipt by the Claimant of the Defendant's letter of response, or (if the Claimant intends to respond to the Counterclaim) after receipt by the Defendant of the Claimant's letter of response to the Counterclaim, the parties should normally meet.

9.2 It is not intended by this Protocol to prescribe in detail the manner in which the meeting should be conducted. However, the Court will normally expect that those attending will include:

9.2.1 where the party is an individual, that individual, and where the party is a corporate body, a representative of that body who has authority to settle or recommend settlement of the dispute;

9.2.2 a legal representative of each party (if one has been instructed);

9.2.3 where the involvement of insurers has been disclosed, a representative of the insurer (who may be its legal representative); and

9.2.4 where a claim is made or defended on behalf of some other party (such as, for example, a claim made by a main contractor pursuant to a contractual obligation to pass on subcontractor claims), the party on whose behalf the claim is made or defended and/or his legal representatives.

9.3 Generally, the aim of the meeting is for the parties to agree what are the main issues in the case, to identify the root cause of disagreement, and to consider (i) whether, and if so how, the case might be resolved without recourse to litigation, and (ii) if litigation is unavoidable, what steps should be taken to ensure that it is conducted in accordance with the overriding objective as defined in rule 1.1 of the Civil Procedure Rules. Alternatively, the meeting can itself take the form of an ADR process such as mediation.

9.4 If the parties are unable to agree on a means of resolving the dispute other than by litigation they should seek to agree:

9.4.1 if there is any area where expert evidence is likely to be required, how expert evidence is to be dealt with including whether a joint expert might be appointed, and if so, who that should be; and (so far as is practicable);

9.4.2 the extent of disclosure of documents with a view to saving costs and to the use of the e-disclosure protocol; and

9.4.3 the conduct of the litigation with the aim of minimising cost and delay.

9.5Any party who attended any pre-action meeting shall be at liberty and may be required to disclose to the Court:

9.5.1 that the meeting took place, when and who attended;

9.5.2 the identity of any party who refused to attend, and the grounds for such refusal;

9.5.3 if the meeting did not take place, why not;

9.5.4 any agreements concluded between the parties; and

9.5.5 the fact of whether alternative means of resolving the dispute were considered or agreed.

9.6 Except as provided in paragraph 9.5, everything said at a pre-action meeting shall be treated as "without prejudice".

10 Other Matters

***10.1** The parties may agree longer periods of time for compliance with any of the steps described above save that no extension in respect of any step shall exceed 28 days in the aggregate.*

***10.2** The Protocol process will be concluded at the completion of the pre-action meeting or, if no meeting takes place, 14 days after the expiry of the period in which the meeting should otherwise have taken place.*

11 Protocol Referee Procedure

***11.1** For the purposes of assisting the parties in participating in and* *complying with the Protocol, the parties may agree to engage in the current version of the Protocol Referee Procedure.*

11.2 The Protocol Referee Procedure shall be published from time to time jointly by TeCSA and TECBAR on their respective websites.

12 Limitation of Action

12.1 If by reason of complying with any part of this protocol a Claimant's claim may be time-barred under any provision of the Limitation Act 1980, or any other legislation which imposes a time limit for bringing an action, the Claimant may commence proceedings without complying with this Protocol. In such circumstances, a Claimant who commences proceedings without complying with all, or any part, of this Protocol must apply to the Court on notice for directions as to the timetable and form of procedure to be adopted, at the same time as he requests the Court to issue proceedings. The Court will consider whether to order a stay of the whole or part of the proceedings pending compliance with this Protocol.
Updated: Friday, 17 February 2017

This has been designed to work in a similar way to the other protocols and to promote the early resolution of disputes at the minimum cost and time, however, it need not be used if both parties agree not to use it. The protocol can also be dispensed with if proceedings need to be issued for limitation purposes.

The timeline of the protocol in summary to the usual type of cavity wall claim is as follows:

- The defendant must acknowledge the letter of claim, failing which the claimant will be entitled to commence proceedings without further compliance with the protocol.

- Within 28 days the defendant must either dispute the jurisdiction of the court or state that it is not the correct defendant or it must provide a letter of response containing the information set out in clause 8.5.

- There should normally follow within 21 days a meeting between the parties legal representatives to delineate the issues in the dispute and to agree on how expert evidence should be handled. In a cavity wall case the authors see no reason why this meeting should not take place by telephone. This meeting will normally be 'Without Prejudice' and is not mandatory but parties may be called to explain to the court why it has not taken place.

- The pre-action protocol ends upon conclusion of this meeting or 14 days after any such meeting would have taken place – i.e. 35 days after receipt of the defendant's letter of response or 21 days after the claimant's response to a counter claim from the defendant although it is very unlikely that a counter claim will arise in a cavity wall case.

The protocol requires the claimant to include the following in the letter of claim:

> *Prior to commencing proceedings, the Claimant or his solicitor shall send to each proposed Defendant (if appropriate to his registered address) a copy of a letter of claim which shall contain the following information:*
>
> *7.1.1 the Claimant's full name and address;*
>
> *7.1.2 the full name and address of each proposed Defendant;*
>
> *7.1.3 a brief summary of the claim or claims including (a) a list of principal contractual or statutory provisions relied on (b) a summary of the relief claimed including, where applicable, the monetary value of any claim or claims with a proportionate level of breakdown. The extent of the brief summary should be proportionate to the claim. Generally, it is not expected or required that expert reports should be provided but, in cases where they are*

> *succinct and central to the claim, they can form a helpful way of explaining the Claimant's position;*
>
> *7.1.4 the names of any experts already instructed by the Claimant on whose evidence he intends to rely, identifying the issues to which that evidence will be directed; and*
>
> *7.1.5 the Claimant's confirmation as to whether or not it wishes the Protocol Referee Procedure to apply as provided below.*

At the outset of the claim the solicitor dealing with the claim should have a basic report from a surveyor confirming the presence of cavity wall insulation and an opinion that it is causing disrepair to the claimant's house, together with an opinion supporting causation between the insulation and the disrepair apparent in the house. It would follow from the combined effect of those two aspects of the report that the defendant installer would be in breach of duty either for installing inappropriate insulation, installing insulation in an ineffective and negligent way and/or by installing the insulation in a property which was not suitable for cavity wall insulation as set out in the previous chapter.

Based on this report the claimant should be asked for more or less the same information as set out in chapter 4 above, adapted slightly to take into account that the claimant is the householder and the defendant is the installer and/or the surveyor.

The normal CPR provisions apply in relation to pre-action disclosure and the claimant will want to see by way of disclosure everything connected with the installation of the insulation including the pre-installation site survey and report, any correspondence between the defendant and the pre-installation surveyor, the method statement for the works and the sign-off sheets etc. and post-installation paperwork.

The expert evidence is likely to be of the same order as in a housing disrepair claim with one surveyor on each side and a Scott Schedule listing the items of disrepair and with each expert giving their opinion in relation to causation and breach of duty. Ultimately, of course, as in any claim the ultimate decision as to whether there has been a breach of duty rests with the trial judge. If there is to be oral evidence from the experts this will result in the claim being tracked or re-allocated into the Fast Track.

Experts Duties

Update – Poor Surveyors' Reports

The writers have experienced that some Surveyors are producing reports, where they are not experienced in cavity wall claims. Before instructing an expert it may be important to:-

1. Check the experts *cv* carefully to make sure the expert has appropriate expertise to produce a report.

2. Remind the expert in the letter of instruction the duties under Part 35.

3. Ensure that the expert deals properly and fully with alternative causation in the report. Your letter of instruction will be key.

Part 35 sets out:-

> *Experts – overriding duty to the court*
>
> *35.3*
>
> *(1) It is the duty of experts to help the court on matters within their expertise.*
>
> *(2) This duty overrides any obligation to the person from whom experts have received instructions or by whom they are paid.*
>
> *Contents of report*

35.10

(1) An expert's report must comply with the requirements set out in Practice Direction 35.

(2) At the end of an expert's report there must be a statement that the expert understands and has complied with their duty to the court.

(3) The expert's report must state the substance of all material instructions, whether written or oral, on the basis of which the report was written.

(4) The instructions referred to in paragraph (3) shall not be privileged[(GL)] against disclosure but the court will not, in relation to those instructions –
(a) order disclosure of any specific document; or
(b) permit any questioning in court, other than by the party who instructed the expert, unless it is satisfied that there are reasonable grounds to consider the statement of instructions given under paragraph (3) to be inaccurate or incomplete.

Expert Evidence – General Requirements

2.1 Expert evidence should be the independent product of the expert uninfluenced by the pressures of litigation.

2.2 Experts should assist the court by providing objective, unbiased opinions on matters within their expertise, and should not assume the role of an advocate.

2.3 Experts should consider all material facts, including those which might detract from their opinions.

2.4 Experts should make it clear –
(a) when a question or issue falls outside their expertise; and

(b) when they are not able to reach a definite opinion, for example because they have insufficient information.

2.5 If, after producing a report, an expert's view changes on any material matter, such change of view should be communicated to all the parties without delay, and when appropriate to the court.

2.6

(1) In a soft tissue injury claim, where permission is given for a fixed cost medical report, the first report must be obtained from an accredited medical expert selected via the MedCo Portal (website at: www.medco.org.uk).

(2) The cost of obtaining a further report from an expert not listed in rule 35.4(3C)(a) to (d) is not subject to rules 45.19(2A)(b) or 45.29I(2A)(b), but the use of that expert and the cost must be justified.

(3) 'Accredited medical expert', 'fixed cost medical report', 'MedCo', and 'soft tissue injury claim' have the same meaning as in paragraph 1.1(A1), (10A), (12A) and (16A), respectively, of the RTA Protocol.

Form and Content of an Expert's Report

3.1 An expert's report should be addressed to the court and not to the party from whom the expert has received instructions.

3.2 An expert's report must:

(1) give details of the expert's qualifications;

(2) give details of any literature or other material which has been relied on in making the report;

(3) contain a statement setting out the substance of all facts and instructions which are material to the opinions expressed in the report or upon which those opinions are based;

(4) make clear which of the facts stated in the report are within the expert's own knowledge;

(5) say who carried out any examination, measurement, test or experiment which the expert has used for the report, give the qualifications of that person, and say whether or not the test or experiment has been carried out under the expert's supervision;

(6) where there is a range of opinion on the matters dealt with in the report –
(a) summarise the range of opinions; and
(b) give reasons for the expert's own opinion;

(7) contain a summary of the conclusions reached;

(8) if the expert is not able to give an opinion without qualification, state the qualification; and

(9) contain a statement that the expert –
(a) understands their duty to the court, and has complied with that duty; and
(b) is aware of the requirements of Part 35, this practice direction and the Guidance for the Instruction of Experts in Civil Claims 2014.

3.3 An expert's report must be verified by a statement of truth in the following form –
I confirm that I have made clear which facts and matters referred to in this report are within my own knowledge and which are not. Those that are within my own knowledge I confirm to be true. The opinions I have expressed represent my true and complete professional opinions on the matters to which they refer.

Part 22 deals with statements of truth. Rule 32.14 sets out the consequences of verifying a document containing a false statement without an honest belief in its truth.)

Key Points:-

a) As in all cases, the experts' overriding duty is to help the Court. They are not there to help the parties.

b) The expert must explain the basis of their qualifications to reach the opinion under Part 35.

c) The expert must explain the basis upon which they formed the opinion.

d) The expert must explain what evidence was considered to reach that opinion, i.e. examination of the property.

e) The expert must explain the rationale behind their opinion.

f) The expert must consider the range of possible causes i.e. damp/ mould lifestyle choices by the owners. The claim will be in difficulties at trial if alternative causes are not excluded with reasons as well as having reasons for the finding that the cavity wall is at fault.

g) There will likely be many cases where an expert could reach a view that the issues are not caused by a failed cavity wall – instead, it could be lifestyle choices by the owner.

Defendant's Own Expert?

A very difficult issue the Court is currently having to deal with is the Defendant seeking its own expert. It is useful to review CPR Part 35 in this regard which sets out:-

Court's power to restrict expert evidence

35.4

(1) No party may call an expert or put in evidence an expert's report without the court's permission.

(2) When parties apply for permission they must provide an estimate of the costs of the proposed expert evidence and identify –
(a) the field in which expert evidence is required and the issues which the expert evidence will address; and
(b) where practicable, the name of the proposed expert.

(3) If permission is granted it shall be in relation only to the expert named or the field identified under paragraph (2). The order granting permission may specify the issues which the expert evidence should address.

3) If permission is granted it shall be in relation only to the expert named or the field identified.
(3A) Where a claim has been allocated to the small claims track or the fast track, if permission is given for expert evidence, it will normally be given for evidence from only one expert on a particular issue.

Questions to Experts

6.1 Where a party sends a written question or questions under rule 35.6 direct to an expert, a copy of the questions must, at the same time, be sent to the other party or parties.

6.2 The party or parties instructing the expert must pay any fees charged by that expert for answering questions put under rule 35.6. This does not affect any decision of the court as to the party who is ultimately to bear the expert's fees.

Single joint expert

7 When considering whether to give permission for the parties to rely on expert evidence and whether that evidence should be from a single joint expert the court will take into account all the circumstances in particular, whether:
(a) it is proportionate to have separate experts for each party on a particular issue with reference to –
(i) the amount in dispute;
(ii) the importance to the parties; and
(iii) the complexity of the issue;
(b) the instruction of a single joint expert is likely to assist the parties and the court to resolve the issue more speedily and in a more cost-effective way than separately instructed experts;
(c) expert evidence is to be given on the issue of liability, causation or quantum;
(d) the expert evidence falls within a substantially established area of knowledge which is unlikely to be in dispute or there is likely to be a range of expert opinion;
(e) a party has already instructed an expert on the issue in question and whether or not that was done in compliance with any practice direction or relevant pre-action protocol;
(f) questions put in accordance with rule 35.6 are likely to remove the need for the other party to instruct an expert if one party has already instructed an expert;
(g) questions put to a single joint expert may not conclusively deal with all issues that may require testing prior to trial;
(h) a conference may be required with the legal representatives, experts and other witnesses which may make instruction of a single joint expert impractical; and
(i) a claim to privilege[(GL)] makes the instruction of any expert as a single joint expert inappropriate.

Key Points:-

a) These are likely to be multi track cases so generally the Court may allow each party an expert.

b) The Court may suggest that the parties are required to put questions to the expert and for those questions to be answered before the Court allows additional experts. This may be more proportionate.

c) The Court may also impose a single joint expert. It may be useful to suggest this.

d) If the Defendant obtains permission from their own surveyor it may be important for the Claimant to consider putting Part 35 questions and thereafter a joint report.

e) It is important that any expert instructed is a chartered surveyor and/ or a RICS surveyor. The writers have experienced that some reports do not properly link the damage to the breach of duty

f) It is also important that the reports are not a 'tick box' exercise and that they properly consider the issues of CIGA and BBA guideline in the report (see below).

CHAPTER TEN
CAVITY WALL CLAIMS: REMEDIES, QUANTUM AND COSTS (LIMITATION UPDATE)

Cavity wall claims are brought by way of claims for breach of contract and in negligence. The measure of damages for a breach of contract claim is for the claimant to be put in the position had the contract been performed properly and in tort to be put back in the position had the negligence not happened.

It can be understood from the foregoing discussion on the potential breaches of duty by installers and the site surveyors they often use to appraise houses for insulation, that if the house was unsuitable for having cavity wall insulation installed then the main cause of action will be negligence and the remedy will be for the extraction of the insulation and repair and/or redecoration of any parts of the property affected by the inappropriate installation. There would also be a claim for breach of contract under s.13 of the Supply of Goods and Services Act 1982 for installations prior to October 2015 for failing to use reasonable care and skill in deciding to go ahead with an installation which should not have gone ahead.

For installations from October 2015 the claim would be under the Consumer Rights Act 2015 for breach of the following:

- Section 9 – Goods to be of satisfactory quality.
- Section 10 – Goods to be fit for particular purpose.
- Section 11 – Goods to be as described.
- Section 49 – Services to be performed with reasonable skill and care.

The cost of extracting cavity wall insulation will depend upon the type of insulation and the size of the property but a typical 3 bedroom semi-detached house is likely to cost between £2,500.00 and £3,000.00 to extract the insulation. Smaller properties or where the extraction is limited could cost as little as £1,200.00.

Where the property was suitable for cavity wall insulation but this was carried out negligently or without due care and skill resulting in the disrepair the same provisions will apply and also s.4 of the 1982 Act where an inappropriate material was used. In such cases the installation can be carried out again but this time in an appropriate manner and with appropriate materials but only after all disrepair has been made good and the property and cavity has dried out sufficiently.

The costs involved in a reinstallation case will naturally be higher than in a case of extraction only and will depend on the precise circumstances of the job. The parties surveyors will include an estimate of such costs in the Scott Schedule and the claimant's surveyor should indicate such costs as part of his initial report unless that report is a preliminary breach of duty indicator only report.

In addition to the cost of cure for remedying the defective installation the claimant will be entitled to damages for the effects of living in a house with damp, mould, cold and the other attendant problems caused thereby. There are currently no reported cases as to quantum for such claims that have come to the authors' attention but such damages are likely to be very similar to those awarded in housing disrepair claims and readers are referred to chapter 5 above for examples of these and the basis on which the courts assess damages.

It is submitted that the measure of damages in cavity wall cases should be by reference to what the house would have rented for had it been let out less the likely diminution of value of the rent caused by the effects of the defective insulation very much as in a housing disrepair case. The analogy seems compelling with the installer taking the place of the landlord in a housing disrepair case.

There would be no difference at all in terms of damages in relation to any personal injury claim arising out of defective cavity wall insulation and if such a claim exceeded £1,000.00 then the claim would be allocated to the Fast Track.

In chapter 9 above it was discussed that many cavity wall claims could fall into the Small Claims Track. In such cases, the costs recoverable from the defendant would be limited to the small claims fixed costs limited to £260.00 plus VAT. Accordingly, such claims are only viable if the claimant is willing to enter into a Damages Based Agreement, (a 'DBA'), at a level that is commercial for the solicitors.

For claims where damages of all kinds are likely to exceed £10,000.00 standard basis costs will apply and as with housing disrepair claims QOCS protection will only attach to any personal injury claim. There will be no QOCS protection for the claims in tort and contract for extraction, repair and reinstatement. For this reason, claimants will need to obtain ATE cover if their claim includes any element of disrepair.

Limitation Update 2017

The following statutory provisions are likely to apply as to limitation:-

> *S2 Limitation Act 1980 Time limit for actions founded on tort.*
>
> ***An action founded on tort shall not be brought after the expiration of six years from the date on which the cause of action accrued.***
>
> *S5 Limitation Act 1980 Time-limit for actions founded on simple contract.*
>
> ***An action founded on simple contract shall not be brought after the expiration of six years from the date on which the cause of action accrued.***

The general rule in contract is that the cause of action accrues when the breach occurs, not when the damage is suffered. In tort, the cause of action accrues when the damage is first sustained. The cause of action, whether in tort or contract, arises regardless of whether or not the Claimant could have known about the damage.

There is no general statutory provision for the extension of time in contract cases when the breach is not immediately discoverable. The cause of action in contract accrued here is 6 years from the date of breach. The contract was breached when the installation was carried out but it does not run from the date of knowledge in contract.

> *14A and 14B of the Limitation Act 1980* may help as to tort claims which sets out:-
>
> ***A Special time limit for negligence actions where facts relevant to cause of action are not known at date of accrual.***
>
> *(1) This section applies to any action for damages for negligence, other than one to which section 11 of this Act applies, where the starting date for reckoning the period of limitation under subsection (4)(b) below falls after the date on which the cause of action accrued.*
>
> *(2) Section 2 of this Act shall not apply to an action to which this section applies.*
>
> *(3) An action to which this section applies shall not be brought after the expiration of the period applicable in accordance with subsection (4) below.*
>
> *(4) That period is either—*
> *(a) six years from the date on which the cause of action accrued; or*
> ***(b) three years from the starting date as defined by subsection (5) below, if that period expires later than the period mentioned in paragraph (a) above.***
>
> ***(5) For the purposes of this section, the starting date for reckoning the period of limitation under subsection (4)(b) above is***

the earliest date on which the plaintiff or any person in whom the cause of action was vested before him first had both the knowledge required for bringing an action for damages in respect of the relevant damage and a right to bring such an action.

(6) In subsection (5) above "the knowledge required for bringing an action for damages in respect of the relevant damage" means knowledge both—
(a) of the material facts about the damage in respect of which damages are claimed; and
(b) of the other facts relevant to the current action mentioned in subsection (8) below.

(7) For the purposes of subsection (6)(a) above, the material facts about the damage are such facts about the damage as would lead a reasonable person who had suffered such damage to consider it sufficiently serious to justify his instituting proceedings for damages against a defendant who did not dispute liability and was able to satisfy a judgment.

(8) The other facts referred to in subsection (6)(b) above are—
(a) that the damage was attributable in whole or in part to the act or omission which is alleged to constitute negligence; and
(b) the identity of the defendant; and
(c) if it is alleged that the act or omission was that of a person other than the defendant, the identity of that person and the additional facts supporting the bringing of an action against the defendant.

(9) Knowledge that any acts or omissions did or did not, as a matter of law, involve negligence is irrelevant for the purposes of subsection (5) above.

(10) For the purposes of this section a person's knowledge includes knowledge which he might reasonably have been expected to acquire —

> *(a) from facts observable or ascertainable by him; or*
> *(b) from facts ascertainable by him with the help of appropriate expert advice which it is reasonable for him to seek;*
> *but a person shall not be taken by virtue of this subsection to have knowledge of a fact ascertainable only with the help of expert advice so long as he has taken all reasonable steps to obtain (and, where appropriate, to act on) that advice.*
>
> ***14B Overriding time limit for negligence actions not involving personal injuries.***
>
> *(1) An action for damages for negligence, other than one to which section 11 of this Act applies, shall not be brought after the expiration of fifteen years from the date (or, if more than one, from the last of the dates) on which there occurred any act or omission—*
> *(a) which is alleged to constitute negligence; and*
> *(b) to which the damage in respect of which damages are claimed is alleged to be attributable (in whole or in part).*
>
> *(2) This section bars the right of action in a case to which subsection (1) above applies notwithstanding that—*
> *(a) the cause of action has not yet accrued; or*
> *(b) where section 14A of this Act applies to the action, the date which is for the purposes of that section the starting date for reckoning the period mentioned in subsection (4)(b) of that section has not yet occurred;*
> *before the end of the period of limitation prescribed by this section.*

Latent Damages Act 1986 (section 1) may also be relevant. Limitation can be a defence and must be pleaded by the Defendant (see *Clerk and Lindsell on Torts at pg 1710 and pg 366*). The burden of proving limitation falls within the limitation period is upon the claimant (*see Cartledge v E Jopling and Sons [1963] AC at 784).* Whenever limitation is raised the Claimant MUST state a positive case on their date of knowledge in a Reply to Defence if not already dealt with pre-emptively in the Particulars of Claim.

CHAPTER ELEVEN
CAVITY WALL CLAIMS AND PERSONAL INJURY CLAIMS (NEW)

Personal Injury Claims can arise out of cavity wall claims. These will be mainly limited to respiratory conditions caused by damp and mould, and could be brought in negligence of breach of contract (if the claimant was a party to the contract). Limitation is likely to be 3 years from date of earliest knowledge, (unless a minor), so the date of knowledge will need to be checked carefully. The 14th edition of the JC Guidelines set out the awards for asthma and respiratory related conditions. There are currently no reported cases for respiratory conditions caused by a failed cavity wall but the following principles are likely to be akin to housing disrepair cases, which can cause similar issues:-

a) It will be important to obtain a report from a respiratory consultant and not just a GP.

b) The personal injury claim will probably commence in the PL portal, as it is technically a public liability claim.

c) QWOCS will apply to the personal injury claim, but <u>not</u> the cavity wall claim, hence ATE insurance is required.

d) The personal injury claim and the cavity wall claims should, arguably, be issued together.

e) It will be important on causation to link any damp / mould in the property to the failed CWI and the respiratory condition.

CHAPTER TWELVE
CHOOSING THE CORRECT DEFENDANT, INSURANCE ISSUES AND INSOLVENT DEFENDANTS

A major issue with cavity wall claims can be that the correct defendant can be difficult to determine. The writers have found that some of the installers are now insolvent and there is no longer any practical redress. Further, even where the insolvent defendant's insurer can be traced, some public liability policies do not cover the rectification works to the property or have a large excess. The writers have encountered large tranches of claims where there is an insurer in the background but where, upon further investigation, the policy simply does not cover the heads of loss being claimed. There is also considerable doubt whether some of the solvent installers will have sufficient funds to meet a claim.

Here is a ten-point checklist for CFA claims:-

1. Ensure that the Defendant is solvent. It is worthwhile conducting a companies house check.

2. If the Defendant is solvent but has limited funds – is the case worth pursuing?

3. If the Defendant is insolvent – is there a public liability or professional indemnity policy that may cover it?

4. If you have a solvent defendant or a PL insurer – consider a PAD application to obtain a copy of the policy to see what it covers. The case law is not fully supportive of obtaining insurance policies as it can be argued that the ability of a defendant to pay the damages sought is not relevant to the court but the writers have had success in this area.

5. If the policy does not cover third party repair / redecoration

work – is it likely the remaining damages are worth pursuing?

6. Does the policy have an excess? If so, what is the client likely to recover?

7. If it is a PII policy – are claims made on a 'claims made' basis – i.e. the policy will not indemnify claims made once the defendant has gone bust?

8. Can you obtain the insurance details from CIGA?

9. If you cannot help the claimant - would the CIGA guarantee cover the works?

10. Are you issuing lots of claims against one company? If the company has no insurance – how many claims can they stand before going bust?

CHAPTER THIRTEEN
COMMON DEFENCES AND CASE MANAGEMENT ISSUES

Cavity Wall Claims Litigation is still in its very early stages, and the writers are not yet aware of any claims that have proceeded to Trial (there are no reported cases). The main defences being pursued appear to be at this stage as follows:-

1. The damage was caused by lifestyle issues caused by the homeowner. The defendant will need to produce a report in this regard and / or put Part 35 questions to the claimant's expert. These issues will also need to be addressed by the claimant's surveyor.

2. The damage was caused by other factors i.e. failed damp proof course / other building works to the property. This as pointed out in a recent High Court hearing, is a matter for the surveyors to consider.

3. There are limitation issues in contract and/ or tort. This is dealt with in other chapters.

4. The damage is pure economic loss, if out of time in contract. It is submitted that the claimant is entitled to recover the damage to the property in tort. These are not claims for pure economic loss, they are claims caused by physical damage to the property. *Clerk and Lindsell at page 338* says "*pure economic loss refers to financial loss suffered by a claimant which* ***is unconnected*** *with,* ***and does not flow from damage to persons or property***". These losses directly flow from damage to property, caused by the Defendant as a result of negligent installation. A claim for increased heating bills, on the other hand, would likely be pure economic loss.

Clerk and Lindsell at page 530 says '*Buildings. At first it was held that builders and architects were the exceptions to Donoghue v Stephenson, so that purchasers and others injured by defects had no remedy against them even if they were negligent. It is now established, however, that reality is in no special position in this respect....* ***Liability under Donoghue v Stevenson began with manufacturers, it is not of course limited to them. It has been extended to negligent designers, installers, erectors and assemblers"***

The writers would point out that these are the preliminary issues the writers are seeing on these cases. One High Court Judge recently pointed out at a CMC that these cases are very much driven by expert evidence and the Court is readily prepared to grant both parties a surveyor's report and call oral evidence if required. It is important to remember that the following will apply to TCC claims and the Court will expect the case management guidance to be strictly adhered to:-

General

60.1

(1) This Part applies to Technology and Construction Court claims ('TCC claims').

(2) In this Part and Practice Direction 60 –
(a) 'TCC claim' means a claim which –
(i) satisfies the requirements of paragraph (3); and
(ii) has been issued in or transferred into the specialist list for such claims;
(b) 'Technology and Construction Court' means any court in which TCC claims are dealt with in accordance with this Part or Practice Direction 60; and
(c) 'TCC judge' means any judge authorised to hear TCC claims.

(3) A claim may be brought as a TCC claim if –
(a) it involves issues or questions which are technically complex; or
(b) a trial by a TCC judge is desirable.

(Practice Direction 60 gives examples of types of claims which it may be appropriate to bring as TCC claims.)

(4) TCC claims include all official referees' business referred to in section 68(1)(a) of the Supreme Court Act 1981[1].

(5) TCC claims will be dealt with:
(a) in a Technology and Construction Court; and
(b) by a TCC judge, unless –
(i) this Part or Practice Direction 60 permits otherwise, or
(ii) a TCC judge directs otherwise.

Specialist list

60.2

(1) TCC claims form a specialist list.

(2) A judge will be appointed to be the judge in charge of the TCC specialist list.

Application of the Civil Procedure Rules

60.3

These Rules and their practice directions apply to TCC claims unless this Part or a practice direction provides otherwise.

Issuing a TCC claim

60.4

A TCC claim must be issued in –
(a) the High Court in London;
(b) a district registry of the High Court; or
(c) a County Court hearing centre specified in Practice Direction 60.

Reply

60.5

Part 15 (Defence and Reply) applies to TCC claims with the modification to rule 15.8 that the claimant must –
(a) file any reply to a defence; and
(b) serve it on all other parties;
within 21 days after service of the defence.

Case management

60.6

(1) All TCC claims are treated as being allocated to the multi-track and Part 26 does not apply.

(2) Part 29 and Practice Direction 29 apply to the case management of TCC claims, except where they are varied by or inconsistent with Practice Direction 60.

Judgments and Orders

60.7

(1) Except for orders made by the court of its own initiative and unless the court otherwise orders, every judgment or order made in claims proceeding in the Technology and Construction Court will be drawn up by the parties, and rule 40.3 is modified accordingly.

(2) An application for a consent order must include a draft of the proposed order signed on behalf of all the parties to whom it relates.

Practice Direction 60 - TCC claims

***2.1** The following are examples of the types of claim which it may be appropriate to bring as TCC claims –*

(a) building or other construction disputes, including claims for the enforcement of the decisions of adjudicators under the Housing Grants, Construction and Regeneration Act 1996;
(b) engineering disputes;
(c) claims by and against engineers, architects, surveyors, accountants and other specialised advisers relating to the services they provide;
(d) claims by and against local authorities relating to their statutory duties concerning the development of land or the construction of buildings;
(e) claims relating to the design, supply and installation of computers, computer software and related network systems;
(f) claims relating to the quality of goods sold or hired, and work done, materials supplied or services rendered;
(g) claims between landlord and tenant for breach of a repairing covenant;
(h) claims between neighbours, owners and occupiers of land in trespass, nuisance etc;
(i) claims relating to the environment (for example, pollution cases);
(j) claims arising out of fires;
(k) claims involving taking of accounts where these are complicated; and
(l) challenges to decisions of arbitrators in construction and engineering disputes including applications for permission to appeal and appeals.

2.2 A claim given as an example in paragraph 2.1 will not be suitable for this specialist list unless it demonstrates the characteristics in rule 60.1(3). Similarly, the examples are not exhaustive and other types of claim may be appropriate to this specialist list.

How to start a TCC claim

3.1 TCC claims must be issued in the High Court or in a County Court hearing centre specified in this practice direction.

3.2 The claim form must be marked in the top right hand corner

'Technology and Construction Court' below the words 'The High Court, Queen's Bench Division' or 'The County Court at _____ '.

3.3 TCC claims brought in the High Court outside London may be issued in any district registry, but it is preferable that wherever possible they should be issued in one of the following district registries, in which a TCC judge will usually be available –
Birmingham, Bristol, Cardiff, Chester, Exeter, Leeds, Liverpool, Manchester, Mold, Newcastle upon Tyne and Nottingham.

3.4 The County Court hearing centres in which a TCC claim may be issued are the following –
Birmingham, Bristol, Cardiff, Central London, Chester, Exeter, Leeds, Liverpool, Manchester, Mold, Newcastle upon Tyne and Nottingham.

Applications before proceedings are issued

4.1 A party who intends to issue a TCC claim must make any application before the claim form is issued to a TCC judge.

4.2 The written evidence in support of such an application must state that the proposed claim is a TCC claim.

Transfer of proceedings

5.1 Where no TCC judge is available to deal with a claim which has been issued in a High Court District Registry or one of the County Court hearing centres listed in paragraph 3.4 above, the claim may be transferred –

(1) if it has been issued in a District Registry, to another District Registry or to the High Court in London; or

(2) if it has been issued in the County Court, to another County Court hearing centre where a TCC judge would be available.

5.2 Paragraph 5.1 is without prejudice to the court's general powers to transfer proceedings under Part 30.
(Rule 30.5(3) provides that an application for the transfer of proceedings to or from a specialist list must be made to a judge dealing with claims in that list.)

5.3 A party applying to a TCC judge to transfer a claim to the TCC specialist list must give notice of the application to the court in which the claim is proceeding, and a TCC judge will not make an order for transfer until he is satisfied that such notice has been given.

Assignment of claim to a TCC judge

6.1 When a TCC claim is issued or an order is made transferring a claim to the TCC specialist list, the court will assign the claim to a named TCC judge ('the assigned TCC judge') who will have the primary responsibility for the case management of that claim.

6.2 All documents relating to the claim must be marked in similar manner to the claim form with the words 'Technology and Construction Court' and the name of the assigned TCC judge.

Applications

7.1 An application should normally be made to the assigned TCC judge. If the assigned TCC judge is not available, or the court gives permission, the application may be made to another TCC judge.

7.2 If an application is urgent and there is no TCC judge available to deal with it, the application may be made to any judge who, if the claim were not a TCC claim, would be authorised to deal with the application.

Case management conference

8.1 The court will fix a case management conference within 14 days of the earliest of these events –

(1) the filing of an acknowledgment of service;

(2) the filing of a defence; or

(3) the date of an order transferring the claim to a TCC.

8.2 When the court notifies the parties of the date and time of the case management conference, it will at the same time send each party a case management information sheet and a case management directions form. (The case management information sheet and the case management directions form are in the form set out in Appendixes A and B to this practice direction.)

8.3 Not less than two days before the case management conference, each party must file and serve on all other parties –

(1) completed copies of the case management information sheet and case management directions form; and

(2) an application notice for any order which that party intends to seek at the case management conference, other than directions referred to in the case management directions form.

8.4 The parties are encouraged to agree directions to propose to the court by reference to the case management directions form.

8.5 If any party fails to file or serve the case management information sheet and the case management directions form by the date specified, the court may –

(1) impose such sanction as it sees fit; and

(2) either proceed with or adjourn the case management conference.

8.6 The directions given at the case management conference will

normally include the fixing of dates for –

(1) any further case management conferences;

(2) a pre-trial review;

(3) the trial of any preliminary issues that it orders to be tried; and

(4) the trial.

Pre-trial review

9.1 When the court fixes the date for a pre-trial review it will send each party a pre-trial review questionnaire. (The pre-trial review questionnaire is in the form set out in Appendix C to this practice direction).

9.2 Each party must file and serve on all other parties completed copies of the questionnaire not less than two days before the date fixed for the pre-trial review.

9.3 The parties are encouraged to agree directions to propose to the court.

9.4 If any party fails to return or exchange the questionnaire by the date specified the court may –

(1) impose such sanction as it sees fit; and

(2) either proceed with or adjourn the pre-trial review.

9.5 At the pre-trial review, the court will give such directions for the conduct of the trial as it sees fit.

Listing

10 *The provisions about listing questionnaires and listing in Part 29 and Practice Direction 29 do not apply to TCC claims.*

Trial

11.1 *Whenever possible the trial of a claim will be heard by the assigned TCC judge.*

11.2 *A TCC claim may be tried at any place where there is a TCC judge available to try the claim.*

CHAPTER FOURTEEN CONCLUSIONS AND THE DBA REGULATIONS 2013

As the authors have set out in previous Chapters some cavity wall claims could run on a DBA. The 2013 Regulations set out:-

2013 No. 000
LEGAL SERVICES, ENGLAND AND WALES
The Damages-Based Agreements Regulations 2013
Made
***
Coming into force
1st April 2013

The Lord Chancellor in exercise of the powers conferred by sections 58AA(4) and (5) and 120(3) of the Courts and Legal Services Act 1990(1), having consulted in accordance with section 58AA(6) of that Act, makes the following Regulations, a draft of which has been laid before and approved by resolution of each House of Parliament in accordance with section 120(4)(2) of that Act.

Citation, commencement, interpretation and application

1.—(1) These Regulations may be cited as the Damages-Based Agreements Regulations 2013 and come into force on 1st April 2013.

(2) In these Regulations—
"the Act" means the Courts and Legal Services Act 1990;
"claim for personal injuries" has the same meaning as in Rule 2.3 of the Civil Procedure Rules 1998(3);
"client" means the person who has instructed the representative to provide advocacy services, litigation services (within section 119 of the Act) or claims management services (within the meaning of section 4(2)(b) of the Compensation Act 2006(4)) and is liable to

make a payment for those services;
"costs" means the total of the representative's time reasonably spent, in respect of the claim or proceedings, multiplied by the reasonable hourly rate of remuneration of the representative;
"employment matter" means a matter that is, or could become, the subject of proceedings before an employment tribunal;
"expenses" means disbursements incurred by the representative, including the expense of obtaining an expert's report and, in an employment matter only, counsel's fees;
"payment" means that part of the sum recovered in respect of the claim or damages awarded that the client agrees to pay the representative, and excludes expenses but includes, in respect of any claim or proceedings to which these regulations apply other than an employment matter, any disbursements incurred by the representative in respect of counsel's fees;
"representative" means the person providing the advocacy services, litigation services or claims management services to which the damages-based agreement relates.

(3) Subject to paragraphs (4), (5) and (6), these Regulations shall apply to all damages-based agreements entered into on or after the date on which these Regulations come into force.

(4) Subject to paragraph (6), these Regulations shall not apply to any damages-based agreement to which section 57 of the Solicitors Act 1974(5) (non-contentious business agreements between solicitor and client) applies.

(5) In these Regulations—
(a) regulation 4 does not apply; and
(b) regulations 5, 6, 7 and 8 only apply,
to any damages-based agreement in respect of an employment matter.

(6) Where these Regulations relate to an employment matter, they apply to all damages-based agreements signed on or after the date on which these Regulations come into force.

Revocation of 2010 Regulations and transitional provision

2.—(1) Subject to paragraph (2), the Damages-Based Agreements Regulations 2010(6) ("the 2010 Regulations") are revoked.

(2) The 2010 Regulations shall continue to have effect in respect of any damages-based agreement to which those Regulations applied and which was signed before the date on which these Regulations come into force.

Requirements of an agreement in respect of all damages-based agreements

3. The requirements prescribed for the purposes of section 58AA(4)(c) of the Act are that the terms and conditions of a damages-based agreement must specify—

(a) the claim or proceedings or parts of them to which the agreement relates;

(b) the circumstances in which the representative's payment, expenses and costs, or part of them, are payable; and

(c) the reason for setting the amount of the payment at the level agreed, which, in an employment matter, shall include having regard to, where appropriate, whether the claim or proceedings is one of several similar claims or proceedings.

Payment in respect of claims or proceedings other than an employment matter

4.—(1) In respect of any claim or proceedings, other than an employment matter, to which these Regulations apply, a damages-based agreement must not require an amount to be paid by the client other than—

(a) the payment, net of—

(I) any costs (including fixed costs under Part 45 of the Civil Procedure Rules 1998); and

(ii) where relevant, any sum in respect of disbursements incurred by the representative in respect of counsel's fees,

that have been paid or are payable by another party to the proceed-

ings by agreement or order; and
(b) any expenses incurred by the representative, net of any amount which has been paid or is payable by another party to the proceedings by agreement or order.

(2) In a claim for personal injuries—
(a) the only sums recovered by the client from which the payment shall be met are—
(i) general damages for pain, suffering and loss of amenity; and
(ii) damages for pecuniary loss other than future pecuniary loss,
net of any sums recoverable by the Compensation Recovery Unit of the Department for Work and Pensions; and
(b) **subject to paragraph (4), a damages-based agreement must not provide for a payment above an amount which, including VAT, is equal to 25% of the combined sums in paragraph (2)(a)(i) and (ii) which are ultimately recovered by the client.**

(3) Subject to paragraph (4), in any other claim or proceedings to which this regulation applies, a damages-based agreement must not provide for a payment above an amount which, including VAT, is equal to 50% of the sums ultimately recovered by the client.

(4) The amounts prescribed in paragraphs (2)(b) and (3) shall only apply to claims or proceedings at first instance.

Information required to be given before an agreement is made in an employment matter

5.—(1) In an employment matter, the requirements prescribed for the purposes of section 58AA(4)(d) of the Act are to provide—
(a) information to the client in writing about the matters in paragraph (2); and
(b) such further explanation, advice or other information about any of those matters as the client may request.
(2) Those matters are—
(a) the circumstances in which the client may seek a review of costs

and expenses of the representative and the procedure for doing so;
(b) the dispute resolution service provided by the Advisory, Conciliation and Arbitration Service (ACAS) in regard to actual and potential claims;
(c) whether other methods of pursuing the claim or financing the proceedings, including—
(I) advice under the Community Legal Service,
(ii) legal expenses insurance,
(iii) pro bono representation, or
(iv) trade union representation,
are available, and, if so, how they apply to the client and the claim or proceedings in question; and
(d) the point at which expenses become payable; and
(e) a reasonable estimate of the amount that is likely to be spent upon expenses, inclusive of VAT.

Additional causes of action in an employment matter

6. In an employment matter, any amendment to a damages-based agreement to cover additional causes of action must be in writing and signed by the client and the representative.

Payment in an employment matter

7. In an employment matter, a damages-based agreement must not provide for a payment above an amount which, including VAT, is equal to 35% of the sums ultimately recovered by the client in the claim or proceedings.

Terms and conditions of termination in an employment matter

8.—(1) In an employment matter, the additional requirements prescribed for the purposes of section 58AA(4)(c) of the Act are that the terms and conditions of a damages-based agreement must be in accordance with paragraphs (2), (3) and (4).

(2) If the agreement is terminated, the representatives may not

charge the client more than the representative's costs and expenses for the work undertaken in respect of the client's claim or proceedings.

(3) The client may not terminate the agreement—
(a) after settlement has been agreed; or
(b) within seven days before the start of the tribunal hearing.

(4) The representative may not terminate the agreement and charge costs unless the client has behaved or is behaving unreasonably.

(5) Paragraphs (3) and (4) are without prejudice to any right of either party under general law of contract to terminate the agreement.

Signed by authority of the Lord Chancellor
Parliamentary Under Secretary of State
Ministry of Justice
Date

EXPLANATORY NOTE
(This note is not part of the Order)

These Regulations prescribe the requirements with which a damages-based agreement ("DBA") must comply in order to be enforceable under section 58AA of the Courts and Legal Services Act 1990 (c.41) ("the Act").

DBAs are a type of 'no win, no fee' agreement under which a representative (defined in these Regulations as a person providing the advocacy services, litigation services or claims management services to which the DBA relates) can recover an agreed percentage of a client's damages if the case is won ("the payment"), but will receive nothing if the case is lost.

Prior to amendment by the Legal Aid, Sentencing and Punishment of Offenders Act 2012 (c.10), section 58AA of the Act only provided for the regulation of DBAs used in employment matters. The effect of the amendment, subject to exceptions, is to permit and regulate the use of DBAs in all civil litigation.

These Regulations apply to all DBAs, including those which relate to employment matters, entered into or signed on or after the date on which they come into force. However, section 58AA(9) of the Act provides that, where section 57 of the Solicitors Act 1974 (c.47) applies to a DBA (other than one relating to an employment matter), it is not unenforceable only because it does not satisfy the conditions in section 58AA(4), under which these Regulations are made. Accordingly, article 1(4) excludes those DBAs to which section 57 of the Solicitors Act 1974 applies from the scope of these Regulations. Further, the effect of regulation 4 is limited to all DBAs other than those which relate to employment matters, whereas regulations 5, 6, 7 and 8 only apply to DBAs in respect of employment matters.

Regulation 2 revokes the Damages-Based Agreements Regulations 2010 (SI 2010/1206), which applied only to employment matters and which will continue to have effect in respect of any DBA relating to an employment matter signed before the date on which these Regulations come into force. However, these Regulations make similar separate provision in respect of employment matters.

Regulation 3 applies to all DBAs and specifies the requirements of a DBA.

Regulation 4, which applies to all DBAs other than those which relate to employment matters, **provides that the payment from a client's damages shall be the sum agreed to be paid (which, where relevant, will include any disbursements incurred by the representative in respect of counsel's fees) net of any costs (including fixed costs), or sum in respect of counsel's fees, payable to the representative by another party to the proceedings.**

Regulation 4 also provides that:

> ***in a claim for personal injuries, the amount to be paid by a client, including VAT, must not be greater than 25% of the combined total of the damages recovered by the client in the proceedings for pain, suffering and loss of amenity and pecuniary loss (other than future pecuniary loss), net of any sums recoverable by the Compensation Recovery Unit. The 25% cap will only apply to claims or proceedings at first instance; and in any other claim or proceedings to which these Regulations apply, the amount of the payment, including VAT, must not be greater than 50% of the sums ultimately recovered by the client. The 50% cap will only apply to claims or proceedings at first instance.***

Regulations 5, 6, 7 and 8 apply only to DBAs in respect of employment matters.

Regulation 5 specifies the information that a representative must provide before a DBA is made.

Regulation 6 specifies that additional causes of action can be added to the agreement by written and signed amendment.
Regulation 7 provides for the maximum amount that is payable to the representative from a client's damages under a DBA in respect of an employment matter, so that the amount of the payment, including VAT, must not be greater than 35% of the sums ultimately recovered by the client in the claim or proceedings.

Regulation 8 states that the terms and conditions of an agreement that provide for the termination of the DBA in an employment matter must comply with the following requirements: if the agreement is ended then the representative cannot charge more than his or her costs and expenses for the work done in respect of the client's claim or proceedings; the client may not end the agreement at particular stages; the representative may not end the agreement unless the client has been or is being unreasonable; nothing in regulation 8 prevents a party from exercising a right under the general law of contract to terminate the agreement, for example for misrep-

resentation or fundamental breach.

Key Points:-

a) Claims may be better pursued on a hybrid DBA/CFA. If the claim exceeds £10,000 costs would be recoverable. If the claim does not exceed £10,000, the client can be asked to 50% including vat of the damages on a DBA. It is worth instructing cost Counsel to draft the retainer.

b) The amount of the payment, including VAT, must not be greater than 50% of the sums ultimately recovered by the client. If it is, the retainer will be invalid.

c) In a claim for personal injuries, the amount to be paid by a client, including VAT, must not be greater than 25% of the combined total of the damages recovered by the client in the proceedings for pain, suffering and loss of amenity and pecuniary loss (other than future pecuniary loss), net of any sums recoverable by the Compensation Recovery Unit. The 25% cap will only apply to claims or proceedings at first instance.

d) The agreement must comply with Regulation 3 as above.

e) Cavity wall claims could be over £10,000 and could include heads of loss such as cost of scaffolding, cavity clearance, removal of render, brickwork replacement, reinstallation of CWI, rendering, skips and the costs of the building site, as well as internal costs of decoration. Until the full surveyor's report or schedule of remediation work has been undertaken, it is difficult to know whether the claim will exceed £10,000 and this a hybrid CFA/DBA may be better in the first instance, to offer protection in both respects.

f) Any retainer must strictly comply with the 2013 DBA Regulations, or it will not be enforceable.

g) If the claim is clearly over £10,000, a straight-forward a CFA can be used, but the definition of success must be clear in the CFA, i.e. does the definition of success include recovering damages, as well as removal of the failed installation? The same must be considered for housing disrepair cases. It is worth asking Counsel to draft the CFA if in doubt. Any CFA in this area must be risk-assessed in order to justify the uplift being sought even though the uplift is capped and may well be less than the risk-assessed uplift.

MORE BOOKS BY LAW BRIEF PUBLISHING

A selection of our other titles available now:

'A Practical Guide to the General Data Protection Regulation (GDPR)' by Keith Markham
'A Practical Guide to Digital and Social Media Law for Lawyers' by Sherree Westell
'A Practical Guide to Holiday Sickness Claims, 2nd Edition' by Andrew Mckie & Ian Skeate
'A Practical Guide to Inheritance Act Claims by Adult Children Post-Ilott v Blue Cross' by Sheila Hamilton Macdonald
'A Practical Guide to Elderly Law' by Justin Patten
'Arguments and Tactics for Personal Injury and Clinical Negligence Claims' by Dorian Williams
'A Practical Guide to QOCS and Fundamental Dishonesty' by James Bentley
'A Practical Guide to Drone Law' by Rufus Ballaster, Andrew Firman, Eleanor Clot
'Practical Mediation: A Guide for Mediators, Advocates, Advisers, Lawyers, and Students in Civil, Commercial, Business, Property, Workplace, and Employment Cases' by Jonathan Dingle with John Sephton
'Practical Horse Law: A Guide for Owners and Riders' by Brenda Gilligan
'A Comparative Guide to Standard Form Construction and Engineering Contracts' by Jon Close
'A Practical Guide to Compliance for Personal Injury Firms Working With Claims Management Companies' by Paul Bennett
'A Practical Guide to the Landlord and Tenant Act 1954: Commercial Tenancies' by Richard Hayes & David Sawtell
'A Practical Guide to Personal Injury Claims Involving Animals' by Jonathan Hand
'A Practical Guide to Psychiatric Claims in Personal Injury' by Liam Ryan
'Introduction to the Law of Community Care in England and Wales' by Alan Robinson

'A Practical Guide to Dog Law for Owners and Others' by Andrea Pitt
'Ellis and Kevan on Credit Hire, 5th Edition' by Aidan Ellis & Tim Kevan
'RTA Allegations of Fraud in a Post-Jackson Era: The Handbook, 2nd Edition' by Andrew Mckie
'RTA Personal Injury Claims: A Practical Guide Post-Jackson' by Andrew Mckie
'On Experts: CPR35 for Lawyers and Experts' by David Boyle
'An Introduction to Personal Injury Law' by David Boyle
'A Practical Guide to Running Housing Disrepair and Cavity Wall Claims' by Andrew Mckie, Ian Skeate, Simon Redfearn
'A Practical Guide to Claims Arising From Accidents Abroad and Travel Claims' by Andrew Mckie & Ian Skeate
'A Practical Guide to Cosmetic Surgery Claims' by Dr Victoria Handley
'A Practical Guide to Chronic Pain Claims' by Pankaj Madan
'A Practical Guide to Claims Arising from Fatal Accidents' by James Patience
'A Practical Approach to Clinical Negligence Post-Jackson' by Geoffrey Simpson-Scott
'A Practical Guide to Personal Injury Trusts' by Alan Robinson
'Occupiers, Highways and Defective Premises Claims: A Practical Guide Post-Jackson' by Andrew Mckie
'Employers' Liability Claims: A Practical Guide Post-Jackson' by Andrew Mckie
'A Practical Guide to Subtle Brain Injury Claims' by Pankaj Madan
'The Law of Driverless Cars: An Introduction' by Alex Glassbrook
'A Practical Guide to Costs in Personal Injury Cases' by Matthew Hoe
'A Practical Guide to Alternative Dispute Resolution in Personal Injury Claims – Getting the Most Out of ADR Post-Jackson' by Peter Causton, Nichola Evans, James Arrowsmith
'A Practical Guide to Personal Injuries in Sport' by Adam Walker & Patricia Leonard

'A Practical Guide to Marketing for Lawyers' by Catherine Bailey & Jennet Ingram
'The No Nonsense Solicitors' Practice: A Guide To Running Your Firm' by Bettina Brueggemann
'Baby Steps: A Guide to Maternity Leave and Maternity Pay' by Leah Waller
'The Queen's Counsel Lawyer's Omnibus: 20 Years of Cartoons from the Times 1993-2013' by Alex Steuart Williams

www.ingramcontent.com/pod-product-compliance
Lightning Source LLC
LaVergne TN
LVHW052324200926
848481LV00002B/661